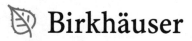

# Hydrological, Geochemical and Geophysical Changes Related to Earthquakes and Slow-Slip Events

Edited by
Chi-Yu King · Michael Manga

Previously published in *Pure and Applied Geophysics* (PAGEOPH),
Volume 175, No. 7, 2018

*Editors*
Chi-Yu King
Earthquake Prediction Research
El Macero, CA, USA

Michael Manga
University of California, Berkeley
Berkeley, CA, USA

ISSN 2504-3625
ISBN 978-3-030-02495-6

Library of Congress Control Number: 2018961206

Cover illustration: Mud volcanoes in the southern Imperial Valley, California are examples of features that respond to earthquakes (photo: May 14, 2014, © Michael Manga)

This book is published under the imprint Birkhäuser, www.birkhauser-science.com by the registered company Springer Nature Switzerland AG
The registered company address is: Gewerbestrasse 11, 6330 Cham, Switzerland

# Contents

Pure Appl. Geophys.
© 2018 Springer International Publishing AG, part of Springer Nature
https://doi.org/10.1007/s00024-018-1923-9

| Pure and Applied Geophysics

# Hydrological, Geochemical and Geophysical Changes Related to Earthquakes and Slow-Slip Events: Introduction

CHI-YU KING[1] and MICHAEL MANGA[2]

It has been documented for thousands of years that earthquakes and other tectonic processes have hydrological effects. Fault ruptures and the seismic waves they produce deform the crust, so it is expected that crustal fluids respond. The magnitude of the hydrological response can be very large because small stresses can be greatly amplified as changes in water pressure. This is also the premise behind searching for earthquake precursors; the very small hypothesized strains leading up to rupture might have hydrological, geochemical or geophysical manifestations. The magnitude and spatial and temporal patterns of reported changes are not always straightforward to explain and hence, remain the subject of active research.

Ten of the 11 papers collected in this special issue present new observations and explanations. They include various earthquake-related hydrological, geochemical and geophysical changes observed in several seismic regions of Japan, Taiwan, Baja California in Mexico, and China mainland. The majority (seven) of these papers are contributed from China, including a laboratory study on rock mechanics and a brief overview of the Chinese efforts in earthquake prediction research over the past half-century. Similar collections of papers were published earlier, and they include King (1980, 1981, 1984, 1986), Dubois (1995) and Perez et al. (2006, 2007, 2008). For a book providing a review of the interactions between earthquakes and water, see Wang and Manga (2010);

a number of other texts provide overviews of how fluid are affected by mechanical deformation of rocks (e.g., Wang 2000; Coussy 2004; Segall 2010).

The most important goal of studying the various earthquake-related changes is to find possible earthquake precursors that might be useful for earthquake forecasting. However, because of the complexity of the real Earth with many confounding environmental and geological variables, many claimed precursory signals turned out not to be precursors to earthquakes. During the past two decades, the mainstream opinion has been that earthquakes are not predictable and the pursuit of prediction should not be continued (Geller et al. 1997). The discovery of slow-slip events provides renewed hope for identifying and understanding precursory signals.

King (2018) describes the characteristics of a sensitive well in Japan that showed many co-seismic and several pre-earthquake water level changes, some of which may originate from slow-slip events. The well is sensitive because it taps a highly permeable aquifer connected to one side of a nearby fault consisting of an impermeable gouge layer sandwiched between two fractured walls and subject to a large hydraulic gradient.

Streamflow changes after earthquakes are some of the most dramatic hydrological responses to earthquakes. Liu et al. (2017) report streamflow changes at 23 gauges in central Taiwan after the 1999 $M_W7.6$ Chi–Chi earthquake. Post-earthquake increases were recorded at 22 gauges and are attributed to rock fracturing by seismic shaking as well as pore pressure rise due to compressive strain. A large post-earthquake decrease was recorded immediately after the earthquake at the gauge located 4 km from the epicenter on the hanging wall of the fault. They attribute

---
[1]  Earthquake-Prediction Research Inc, Los Altos, CA, USA.
E-mail: chiyuking@gmail.com
[2]  Department of Earth and Planetary Science, University of California, Berkeley, CA 94720, USA. E-mail: manga@seismo.berkeley.edu

this decrease to infiltration through the streambed due to a co-seismic decrease in pore pressure induced by co-seismic crustal extension. King and Chia (2017) showed that this 8-month-long streamflow decrease was preceded by a comparable increase starting 4 days before the earthquake. Further, they document a pre-earthquake groundwater level change at a well 1.5 km from the surface trace of the fault. Since both the stream and the well are on the hanging wall, they attribute the pre-earthquake streamflow increase to gravity-driven groundwater discharge into the creek due to crustal buckling, and the water level anomaly to shallow slow-slip events triggered by the post-earthquake downward water flow.

The most abundant data come from wells. Sun et al. (2017) use the Molchan (1990) error diagram test to interpret data from a well in China, and conclude that the well is somewhat predictive of regional earthquakes, with a prediction gain of a factor of two within 30 days of water level anomalies. Ma et al. (2017) document changes in water level after the 2008 Wenchuan earthquake and conclude that changes in permeability can explain the observed water level changes. Zhang et al. (2017) similarly show that earthquake-induced changes in permeability can explain changes in water level near the Three Gorges, China, after that same earthquake. Sarychikhina et al. (2018) compare groundwater level and temperature changes with ground deformation recorded before and after a magnitude 5.4 earthquake and its fore-shock and aftershocks in the Mexicali Valley of Baja California, Mexico. They attribute the co-seismic water level changes to static volumetric strains caused by the main shock, except in one well, where the water level change may have been affected also by a triggered slip event on a nearby fault. Some of the co-seismic temperature changes are explained by convection and mixing of groundwater by seismic shaking. These changes together with some gradual pre-earthquake changes are explainable by the dila-tancy–diffusion theory, or possibly by assuming the occurrence of a slow-slip event and/or fault permeability changes.

Wang et al. (2018a, b) studied the stress evolution during a 10-year period before the 2008 $M_s$ 8.0 Wenchuan earthquake around the seismogenic fault using the method of cataclastic analysis (Rebetsky 2009). They found some significant stress changes before the earthquake.

Shi et al. (2017) interpret groundwater and animal behavior anomalies along a fault zone in the Xichang area, southwestern Sichuan Province, China, from May to June 2002, after which no major earthquake occurred. A comparison with geodetic data and seismicity suggests that the anomalies may be the result of increased tectonic activity in the Sichuan–Yunnan block.

Ren et al. (2017) report a laboratory rock mechanics experiment in which a series of stick–slip events were generated along a precut planar strike slip fault in a granodiorite block under bi-axial compression. Temperature increased prior to each slip event in sections of the fault plane due to partial sliding, but decreased inside the rock itself due to stress relaxation. The regions of increased temperatures expanded rapidly immediately prior to each stick–slip event.

In the absence of quality long-term records, it is not possible to establish the reliability of reported precursory signals. Wang et al. (2018a, b) provide brief overview of the extensive effort since 1966 to monitor hydrological and geochemical changes for the purpose of earthquake prediction in China. Since the results of previous studies using this data were published mostly in Chinese, this overview should help those who cannot read Chinese to understand what data were collected and what has been learned to date.

The papers in this special issue also make clear that to mount an effective effort to search for, identify, and then understand precursory signals, it is important to recognize the heterogeneity of the crust and to deploy appropriate monitoring instruments at "sensitive sites", such as along faults or weak zones and especially on the hanging walls in case of thrust faults, where crustal strain may concentrate. At the same time, models used to interpret data need to account for this geological heterogeneity. In view of rarity of destructive earthquakes in any given region and for the purpose of understanding background variations, it is important to maintain long-term monitoring and data collection.

*Acknowledgements*

We thank the following reviewers for their help in evaluating the submitted manuscripts: P.F. Bari, Y. Chi, G. De Luca, B. Delbridge, Y. Fialko, P. Fulton, Z. Geballe, S. Hao, G.C.P. King, Y. Kitagawa, N. Koizumi, G. Lai, X. Lei, K.F. Ma, G. Li, C. Mohr, Y.L. Rebetsky, Z. Shi, R.P. Singh, Y. Taran, C.-Y. Wang, G. Wang, L. Xue, S. Yabe, R. Yan, L. Yi, Y. Zhang, X. Zhou.

REFERENCES

Coussy, O. (2004). *Poromechanics*. Hoboken: Wiley.

Dubois, C. (1995). *Gas geochemistry*. Northwood: Science Reviews.

Geller, R. J., Jackson, D. D., Kagan, Y. Y., & Mulargia, F. (1997). Earthquakes cannot be predicted. *Science, 275*, 1616.

King, C.-Y. (1980). Geochemical measurements pertinent to earthquake prediction. *Journal of Geophysical Research, 85*, 3051.

King, C.-Y. (1981). A special collection of reports on earthquake prediction: Hydrologic and geochemical studies. *Geophysical Research Letters, 8*, 421–424.

King, C.-Y. (1984) Earthquake hydrology and chemistry. *Pure and Applied Geophysics, 122*, 141–142.

King, C.-Y. (1986). Preface to gas geochemistry of volcanism, earthquakes, resource exploration, and Earth's interior. *Journal of Geophysical Research, 91*, 12157.

King, C. -Y. (2018). Characteristics of a sensitive well showing pre-earthquake water-level changes. *Pure and Applied Geophysics* https://doi.org/10.1007/s00024-018-1855-4.

King, C. -Y., & Chia, Y. (2017). Anomalous streamflow and groundwater-level changes before the 1999 M7.6 Chi-Chi earthquake in Taiwan: Possible mechanisms. *Pure and Applied Geophysics*. https://doi.org/10.1007/s00024-017-1737-1

Liu, C. -Y., Chia, Y., Chuang, P. -Y., Wang, C. -Y, Ge, S., & Teng, M. -H. (2017). Streamflow changes in the vicinity of seismogenic fault after the 1999 Chi-Chi earthquake. *Pure and Applied Geophysics*. https://doi.org/10.1007/s00024-017-1670-3

Ma, Y., Wang, G., & Tao, Y. (2017). Hydrological changes induced by distant earthquakes at the Lujiang well in Anhui, China. *Pure and Applied Geophysics*. https://doi.org/10.1007/s00024-017-1710-z

Molchan, G. M. (1990). Strategies in strong earthquake prediction. *Physics of the Earth and Planetary Interiors, 61*, 84–98.

Perez, N. M., Gurrieri, S., King, C.-Y., & McGee, K. (2006). Introduction to terrestrial fluids, earthquakes and volcanoes: The Hiroshi Wakita Volume I. *Pure and Applied Geophysics, 163*, 629–914.

Perez, N. M., Gurrieri, S., King, C.-Y., & McGee, K. (2007). Introduction to terrestrial fluids, earthquakes and volcanoes: The Hiroshi Wakita Volume II. *Pure and Applied Geophysics, 164*, 2373–2571.

Perez, N. M., Gurrieri, S., King, C.-Y., & McGee, K. (2008). Introduction to terrestrial fluids, earthquakes and volcanoes: The Hiroshi Wakita Volume III. *Pure and Applied Geophysics, 165*, 1–180.

Rebetsky, Yu L. (2009). Estimation of stress values in the method of cataclastic analysis of shear fractures. *Doklady Earth Sciences, 428*, 1202–1207.

Ren, Y., Ma, J., Liu, P., & Chen, S. (2017). Experimental study of thermal field evolution in the short-impending stage before earthquakes. *Pure and Applied Geophysics*. https://doi.org/10.1007/s00024-017-1626-7

Sarychikhina, O., Glowacka, E., González, R. V., & Arreazol, M. F. (2018). Analysis and interpretation of earthquake-related groundwater response and ground deformation: A case study of May 2006 seismic sequence in the Mexicali Valley, Baja California, Mexico. *Pure and Applied Geophysics*. https://doi.org/10.1007/s00024-018-1925-7.

Segall, P. (2010). *Earthquake and volcano deformation*. Princeton: Princeton University Press.

Shi, Z., Wang, G., Liu, C., & Che, Y. (2017). Tectonically induced anomalies without large earthquake occurrences. *Pure and Applied Geophysics*. https://doi.org/10.1007/s00024-017-1596-9

Sun, X., Xiang, Y., Shi, Z., & Wang, B. (2017). Preseismic changes of water temperature in the Yushu well, western China. *Pure and Applied Geophysics*. https://doi.org/10.1007/s00024-017-1579-x

Wang, H. F. (2000). *Theory of linear poroelasticity*. Princeton: Princeton University Press.

Wang, C.-Y., & Manga, M. (2010). *Earthquakes and water*. New York: Springer.

Wang, K., Rebetsky, Y., Ma, S., Feng, X., & Liu, G. (2018). Background stress state before the 2008 Wenchuan Earthquake and the dynamics of the Longmen Shan thrust belt. *Pure and Applied Geophysics* https://doi.org/10.1007/s00024-018-1800-6.

Wang, B., Zhu, C., Yan, L., & Yao, Y. (2018). Hydrogeological and geochemical observations for earthquake prediction research in China: A brief overview. *Pure and Applied Geophysics*. https://doi.org/10.1007/s00024-018-1885-y.

Zhang, S., Shi, Z., Wang, G., & Zhang, Z. (2017). Quantitative assessment of the mechanisms of earthquake-induced groundwater level changes in the MP well, Three Gorges area. *Pure and Applied Geophysics*. https://doi.org/10.1007/s00024-017-1643-6

Pure Appl. Geophys.
© 2018 Springer International Publishing AG, part of Springer Nature
https://doi.org/10.1007/s00024-018-1855-4

**Pure and Applied Geophysics**

# Characteristics of a Sensitive Well Showing Pre-Earthquake Water-Level Changes

CHI-YU KING[1]

*Abstract*—Water-level data recorded at a sensitive well next to a fault in central Japan between 1989 and 1998 showed many coseismic water-level drops and a large (60 cm) and long (6-month) pre-earthquake drop before a rare local earthquake of magnitude 5.8 on 17 March 1997, as well as 5 smaller pre-earthquake drops during a 7-year period prior to this earthquake. The pre-earthquake changes were previously attributed to leakage through the fault-gouge zone caused by small but broad-scaled crustal-stress increments. These increments now seem to be induced by some large slow-slip events. The coseismic changes are attributed to seismic shaking-induced fissures in the adjacent aquitards, in addition to leakage through the fault. The well's high-sensitivity is attributed to its tapping a highly permeable aquifer, which is connected to the fractured side of the fault, and its near-critical condition for leakage, especially during the 7 years before the magnitude 5.8 earthquake.

**Key words:** Sensitive site, water well, water level, aquifer, earthquake, slow-slip event, seismic, geodetic, coseismic change, pre-seismic change, subduction zone, seismogenic fault, Japan, plate.

## 1. Introduction

Many hydrologic and geochemical parameters have been monitored since the 1960s for earthquake studies. Among them, groundwater level in wells has been one of the most studied. It has been found that water levels at certain "sensitive wells" may show coseismic, postseismic, and even pre-earthquake changes, other than seismic oscillation, up to epicentral distances of more than 1000 km. Several mechanisms have been proposed to explain such changes, including earthquake-related permeability changes (e.g., reviews by Wakita 1996; King et al. 2006; Wang and Manga 2014).

In two previous studies, King et al. (1999, 2000) presented a set of water-level data that were continuously recorded between 1989 and 1999 at 16 closely clustered wells within 400 m of the Tsukiyoshi fault, which was under a large hydraulic-pressure difference, in Tono mine, central Japan. From the different responses of these wells to many earthquakes, they were able to address such questions as: Why water-level changes recorded at wells close to each other respond so differently to the same earthquakes? (Existence of a fault in between) Why are some wells so "sensitive" to earthquakes? (Tapping aquifers that are connected to a fault under large hydraulic-pressure gradient) Does the sensitivity of a sensitive well stay constant? (No) What are the mechanisms for such changes? (Water leakage through the fault caused by seismic shaking). They also reported some pre-earthquake changes recorded at a sensitive well and attributed such changes to the occurrences of certain crustal-stress increments. However, "What causes the crustal-stress increments?" remained unanswered.

In the present study, I have given a more detailed examination of this set of data to answer this last question. I propose that the cause of the invoked stress increments may be the occurrence of slow-slip events before the corresponding earthquakes.

## 2. Water-Level Data

Tono mine (triangle, Fig. 1) is located in a relatively stable crustal block bounded by several active faults in central Japan where 4 tectonic plates meet: The Pacific plate in the east subducts westward beneath the North American plate along the Japan trench in the north, and beneath the Philippine Sea plate in the south; the Philippine Sea plate subducts

---
[1] Earthquake-Prediction Research, Inc., Los Altos, CA, USA. E-mail: chiyuking@aol.com

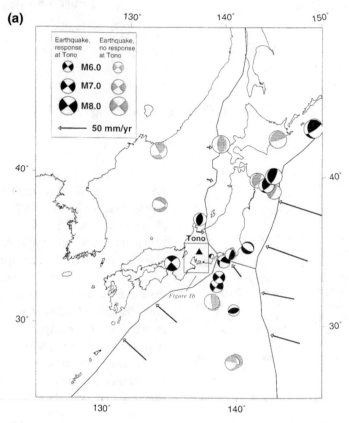

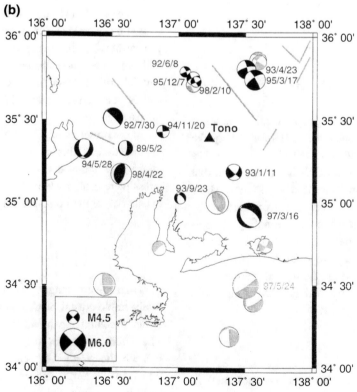

a Tectonic settings in and around Japan, the locations of Tono mine (triangle) in central Japan, and larger earthquakes discussed in the paper (beach balls, darker ones showed earthquake-related changes). In central Japan, the Philippine Sea plate in the south is subducting beneath the Eurasian plate to the northwest and is underlain by the Pacific plate from the east (Ishida 1992), while in northeastern Japan the Pacific plate in the east is subducting beneath the North American plate to the west. The rectangular local area is enlarged in Fig. 1b. b Locations of local earthquakes (beach balls, darker ones showed earthquake-related changes) and active faults (shaded lines) (after King et al. 1999)

northeastward beneath the North American plate along the Sagami trough and northwestward beneath the Eurasian plate along the Nankai trough. Figure 1a shows large earthquakes in Japan (beachballs), and the enlarged map in Fig. 1b shows moderate local earthquakes, during the study period, including a rare local earthquake of magnitude-5.8 on March 16, 1997, the magnitude-7.2 Kobe earthquake in 1995, and a magnitude-8.1 earthquake in 1994 near Hokkaido.

Figure 2 shows a detailed map of the Tono site. The sensitive well (SN3) is located about 300 m north of the Tsukiyoshi fault (dotted line with triangles), which is a reverse fault, dipping about 60° south, away from the well. On the south side of the fault, groundwater was being continuously pumped out of a 126-m deep underground gallery (dotted straight double lines) at a rate of about 50 tons/day to keep the gallery dry. This pumping activity caused a large hydraulic-pressure drop of about 19 kPa on the south side of the fault.

Figure 3 shows coseismic water- level changes recorded at SN3 for 21 out of 45 large distant earthquakes and moderate nearby earthquakes (after the barometric-pressure effect is filtered out for the latter). From Fig. 7, it may be seen that the earthquakes that caused the coseismic water-level changes usually have larger calculated volumetric strain changes at the well than those that did not. All the coseismic changes have about the same shape and duration, i.e., a quick coseismic drop followed by a slower recovery, irrespective of the location, orientation, and focal mechanism of the earthquakes. By comparing these changes with those recorded at the other wells in Tono, King et al. (1999) concluded that they were due to water leakage through seismic

shaking-caused inter-connected fissures in the normally impermeable fault-gauge zone that was subsequently self-healed.

Figure 4 shows the coseismic water-level changes recorded at SN3 and several other deeper wells (see locations in Fig. 2) for the rare magnitude-5.8 local earthquake, which occurred about 50 km away at a depth of 39 km on March 16, 1997 (Fig. 1). The water level at SN3 showed a large quick coseismic drop followed by a slower recovery, similar in shape to the smaller coseismic changes mentioned above, but the post-earthquake rise continued longer to a much higher level. Several days before the earthquake, monitoring began at two other wells, TH-7 and TH-8, which were deep enough to tap the same aquifer as SN3 had (Toki Granite in Fig. 8). These two 200-m deep wells are located about 140 m and 80 m south and north of the fault, respectively, and in them water-pressure was monitored at five different depths with the help of packers (see Fig. 8). The water-pressure data recorded by TH-8 at the Toki Granite level showed a similar coseismic changes as by SN3, but with faster response (as expected, because TH-8 is located closer to the fault than SN3). However, data recorded by TH-7 for the same aquifer on the other side of the fault showed a completely different pattern: No coseismic drop but a post-earthquake sharp peak superimposed on a gradual rise (see bottom curve in the top panel of Fig. 4), the latter shown at three other deeper levels also (see Fig. 8 for the geological columns of the wells). King et al. (1999) attributed the peak to coseismic inflow of leaked water from the higher-pressure side through the fault, and the gradual rise to water sipping down from shallower layers through vertical fissures caused by seismic shaking, a suggestion also proposed by Wang et al. (2016). The big difference in responses to the same earthquake of the closely clustered wells (within about 400 m) located on different sides of the fault shows the importance of recognizing the heterogeneity of the crust in the considerations of their mechanisms.

Figure 5 shows long-term records at SN3 and a shallower well SN1 nearby for the entire study period on a compressed time scale. It is clear that the post-earthquake increase recovered both the coseismic drop (indicated by the last arrow in the figure) and a

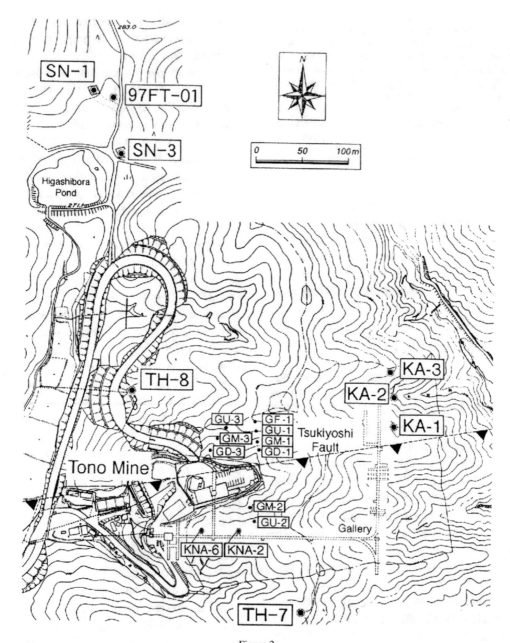

Figure 2
Map of Tono mine area, showing locations of SN1, SN3, TH-7, TH-8, and other wells (circles), the Tsukiyoshi fault (dashed line), underground gallery (dotted lines), the KNA-2 and KNA-2 holes, the topographic contour lines, and roads and various buildings

larger pre-earthquake drop, which occurred about 6 months earlier on 23 September 1996. During this six-month period, water leakage was observed in the gallery, and several attempts were made to stop it, but to no avail. However, the leakage stopped by itself right after the earthquake. This led King et al. (1999) to suggest that the 6-month long water-level drop was caused by earthquake-related leakage through the fault. Subsequently a leakage test was conducted at a fault-crossing hole (KNA2 in Fig. 2) in the gallery, and the result (the large water-level drop in 1998) confirmed this suggestion. There is another large water-level drop in 1995, which was later found to be also due to leakage, caused by equipment-failure at

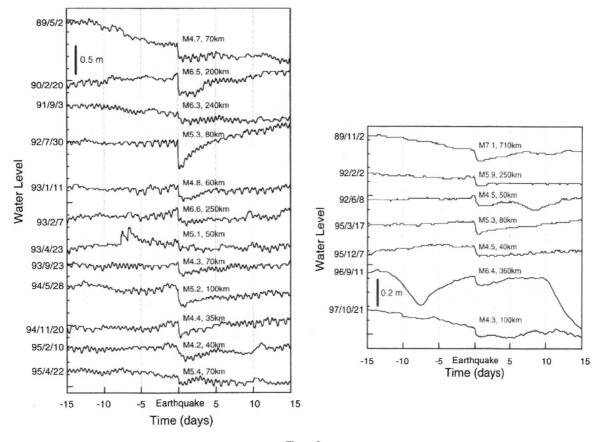

Figure 3
Left panel shows larger coseismic water-level changes recorded at the sensitive well (SN3), together with earthquake dates, magnitudes, and epicentral distances. Right panel shows smaller coseismic changes after correction of barometric-pressure and tidal effects (after King et al. 1999)

another fault-crossing hole, KNA-6 (Fig. 2; King et al. 1999).

The long-term record of SN3 in Fig. 5 also shows some other variations, including 5 water-level drops, each before a large earthquake (indicated by arrow), and subsequent recoveries, as shown on an expanded scale in Fig. 6. These changes are clearly distinguishable from the background noise caused by barometric and tidal variations, which are not filtered out in the data shown in the figure. (Barometric-pressure was measured concurrently with water levels by the same research group which unfortunately has ceased to exist, and there is no archive data with any meteorological organization). These earthquakes were mostly associated with larger calculated volumetric strain changes of about $10^{-7}$ (indicated by diamonds in Fig. 7). They include the magnitude 7.2

Kobe earthquake about 220 km away, the magnitude 8.1 earthquake near Hokkaido about 1200 km away, and a magnitude 6.1 earthquake 510 km away on September 5, 1996.

The water-level drop towards the end of 1998 (Fig. 5) was caused by drilling activity about 1 km away. The drop began about 2 days after a new 1-km hole MIU-2 was drilled through the Tsukiyoshi fault, causing water leakage from the SN3-tapped aquifer through the fault, as discussed in King et al. (2000). Efforts were made to stop the leakage at MIU-2 on the ground surface but to no avail, because water could still leak through the hole outside the casing to other crustal layers. To my knowledge, this has unfortunately destroyed the pressure gradient across the Tsukiyoshi fault, and thus the sensitivity of SN3 well, ever since.

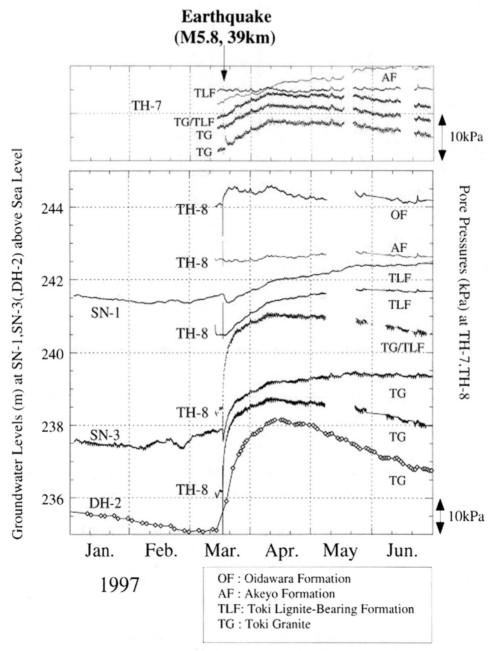

**Earthquake
(M5.8, 39km)**

Groundwater responses at monitoring wells
in the Tono area to a large earthquake

Figure 4
Coseismic water-level changes at SN1, SN3, TH-7 and TH-8 wells for the magnitude 5.8 local earthquake (at time indicated by the arrow)
about 50 km away at a depth of 39 km

As shown in Fig. 6, the above-mentioned pre-earthquake changes are similar in both shape and duration, each consisting of a drop (or slowdown of an increasing trend, for the 17 January 1995 event) beginning about a week before the earthquake, at a rate much smaller than that of the coseismic case.

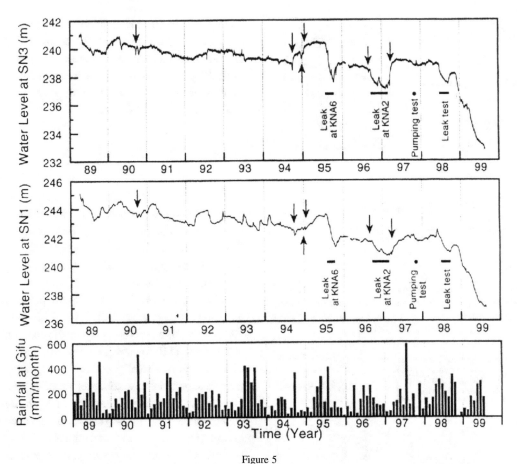

Figure 5

Long-term water-level data recorded at SN1 and SN3 wells during 1989–1999, together with monthly rainfall data. Arrows indicate occurrence of larger earthquakes that showed pre-earthquake water-level changes at SN3 (After King et al. 2000)

The post-earthquake recovery rate is, however, comparable to that of the coseismic drops. Smaller and faster coseismic drops can be seen in the first two curves in Fig. 6. A research group of the Kyoto University also observed coseismic water-level, chemistry and temperature changes elsewhere for the three earthquakes corresponding to the middle three curves in this figure (see King et al. 1995).

The diurnal and semidiurnal variations induced by tidal and barometric- pressure changes are not filtered out in this figure, and they do not seem to show any significant earthquake-related changes in these curves. The possibility of barometric and tidal effects in the water level data before and after the magnitude 5.8 earthquake was checked by King et al. (1999), and a seemingly anomalous change they found turned out not to be really anomalous (See their Fig. 7 and related discussion).

Figure 8 (upper panel) shows the geological columns of the SN3 and other wells. As mentioned above, this well, together with TH-7 and TH-8, taps the high-permeability confined aquifer (the top layer of weathered bed rock, Toki Granite), which intersects the fault zone (lower panel).

The fault zone in this study consists of a fault-gouge layer about 10-30 cm thick, which is weak and normally impermeable, sandwiched in between two highly permeable fractured side layers. Since the aquifer tapped by SN3 intersects the fractured zone on the higher-pressure side, water-level drop at the well could be caused by fissures that occurred anywhere in a wide area of the fault plane.

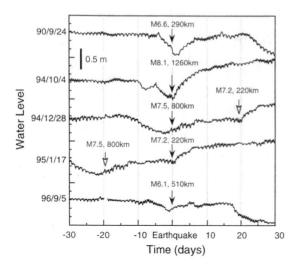

Figure 6

Pre-earthquake water-level changes recorded at SN3 well for 5 large distant earthquakes (arrows) aligned in the middle of a 60-day time window. Earthquake dates are shown on the left side (after King et al. 2000)

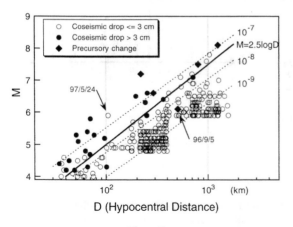

Figure 7

Magnitude (M) as a function of hypocentral distance (D) for the studied earthquakes. Included are all earthquakes of magnitude 5.9 and larger in and near Japan, 4.0–4.8 in the local area shown in Fig. 1b, 4.8 (inclusive) to 5.9 within hypocentral distance of 400 km. The solid line that corresponds to a volumetric strain level of a few $10^{-8}$ separates approximately earthquakes associated with co-seismic changes (larger than 3 cm) at SN3 from those not associated. The diamonds indicate earthquakes associated with pre-earthquake changes, three of five of them around the strain level $10^{-7}$. The dashed lines indicate expected volumetric strain levels from an earthquake model by Dobrovolsky et al. (1979) (after King et al. 1999)

## 3. Discussion

In the following, I will discuss several issues, including: Why was SN3 so sensitive to crustal-stress

increment and seismic shaking? Is there any difference between the permeability changes caused by crustal-stress increments and those by seismic shaking? What is the cause of the stress increments that may have caused the pre-earthquake water-level changes observed in this study?

### 3.1. Sensitivity of SN3 Well

King et al. (1999, 2000) attributed the high-sensitivity of SN3 well to its tapping a highly permeable aquifer, which is connected to the fractured side of the fault. The aquifer/barrier system of this well was apparently in a near-critical condition for water leakage, especially during the 7-year period before the magnitude 5.8 earthquake.

The last earthquake that showed pre-earthquake water-level change at SN3 before this earthquake (Figs. 5 and 6) was a relatively small (magnitude 6.1) and distant (510 km away) event on September 5, 1996, which was shortly before the beginning of the large 6-month long water-level drop. It should not have caused any pre-earthquake or even coseismic change according the magnitude vs. distance distribution (Fig. 7). On the other hand, no significant earthquake-related water-level change was observed after the earthquake for about a year, even when a magnitude 5.9 earthquake occurred only about 100 km away on May 24, 1997 (9 weeks afterwards), as it should according to the same figure. This observation was taken to be an indication of a temporary sensitivity decrease for the well. The sensitivity apparently recovered by February 10, 1998, when a coseismic drop was recorded for a magnitude 4.3 earthquake about 35 km away. Since then a few other coseismic changes were recorded also. These observations led King et al. (1999) to suggest that the local stress might have reached a near-critical level for leakage through the fault several years prior to the magnitude 5.8 earthquake, and then decreased to below this level for about a year after the earthquake.

### 3.2. Stress Increments Caused by Slow-Slip Events

Since the previous study of King et al. (1999, 2000), many slow-slip events have been

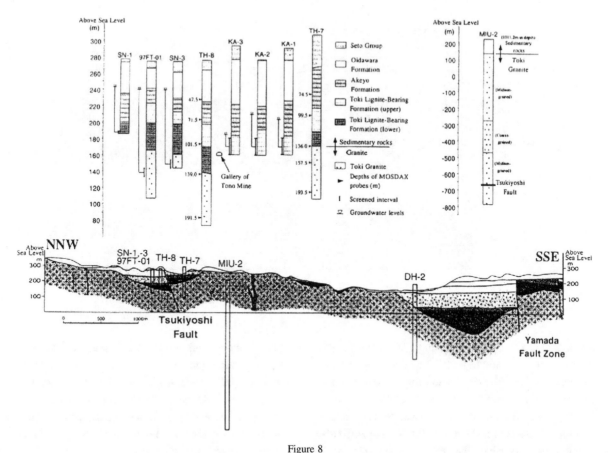

Figure 8
Geological columns of 8 deeper monitoring wells at Tono (upper panel) and a cross-sectional sketch of the local crust (lower panel) (after King et al. 2000)

observed in various subduction zones in the world, including those in Japan. Some of these events have an equivalent magnitude as large as 7, and some may even have triggered large megathrust earthquakes by transferring load to the adjacent seismogenic areas of the plate interface (Linde et al. 1996; Linde and Sacks 2002, 2003; Hirose and Obara 2005; Schwartz and Rokosky 2007; Matsubara et al. 2009; Kato et al. 2012; Uchida et al. 2016; Obara and Kato 2016). Such events may thus be considered as the causes of the stress increments needed to explain the observed pre-earthquake water-level changes in the present study.

Kobayashi (2014) reported the occurrence of a long-term magnitude-6.7 slow-slip event that began in early 1996 and lasted about 1–1.5 years about 20–40 km below the nearby Kii Channel (within 200 km of Tono), in an area of the interface between

the subducting Philippine Sea plate and the Eurasian plate in Nankai trough. This subduction zone is characterized by a seismogenic area about 10–20 km deep located in between two transition areas in the plate interface (Fig. 9), and the one downdip is where this slow-slip event occurred (Matsubara et al. 2009). The earlier part of this event, which probably accounts for most of the slip of the event because such events usually begin rapidly and end slowly (King et al. 1973), may possibly be the cause of a stress increment in the Kii peninsula up to Tono, where the magnitude 6.1 earthquake on 5 September 1996, the 6-month long water-level anomaly at SN3, and the subsequent magnitude 5.8 earthquake on 16 March 1997 occurred. To check this possibility, the static volumetric strain change induced by this event (with an assumed rupture length of 50 km, width 50 km, and a slip of 20 cm) at Tona at a distance of

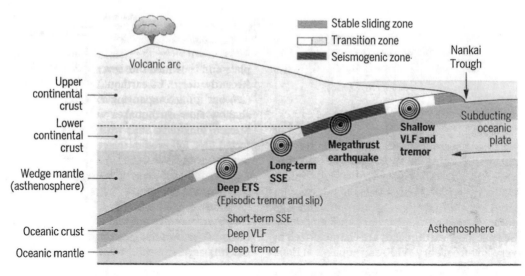

Figure 9
A cross-sectional sketch of the Nankai subduction zone in Japan (after Obara and Kato 2016)

about 200 km is calculated using the Okada (1992) model, and it is found to be $10^{-9}$, which is apparently too small. However, the Okada-model assumes the heterogeneous crust with many faults to be a homogeneous poro-elastic half space. The half-space assumption is obviously not valid for Nankai subduction zone, for which the fault slip at the inter-plate boundary downdip the transition depth of about 40 km is stable sliding (thus compliant). Thus, the model probably has underestimated the amount of load transfer between adjacent sections of the zone after the occurrences of earthquakes and slow-slip event, such as between the Kobe/Kii Channel and the adjacent Kii peninsula area to the east. A better model for the crust would be a 40-km thick plate with free boundaries at the top and the bottom, especially for the slow-slip events, which have large source dimensions as well as slow motion. Similarly, the assumption of homogeneity and poro-elasticity may have underestimated the stress/strain at the Tsu-kiyoshi fault, to which the SN3 well is hydraulically connected, due to the stress concentration effect. For the same reasons, the model cannot explain many other water-level changes triggered at long distances either, as discussed later. This situation is similar to the case of brittle solid in fracture mechanics (Hertzberg 1989) that the actual tensile strength is two orders of magnitude smaller than the theoretical

strength calculated on the assumption of its being perfectly homogeneous and elastic. The reason for the discrepancy is that such solid is not perfect, as assumed, but always contains micro-cracks, where the applied stress is greatly concentrated/amplified and fracture occurs at the weakest of them at the actually observed lower level.

Alternatively, the volumetric strain increment at Tono induced by this magnitude-6.7 event can be estimated using a model by Dobrovolsky et al. (1979). As shown in Fig. 7, at a distance of 200 km, the estimated strain increment is about $5 \times 10^{-7}$, which is about the same as that of the magnitude 7.2 Kobe earthquake (a diamond in Fig. 7) that occurred on 17 January 1995 near the same zone (Nankai Trough), far exceeding the threshold of the earthquake vs. water-level relatedness (solid line in Fig. 7) that was empirically determined from the water-level data recorded at SN3 in this study. This strain increment also exceeds the threshold of most observed earthquake-related hydrological changes summarized in Wang and Manga (2014). Comparable strain changes may have also facilitated the occurrence of the rare local magnitude 5.8 earthquake on March 16 and another on May 24, 1997 (Fig. 1b). The model of Dobrovolsky et al. (1979) deals with deformations at the Earth's surface from the appearance of a soft inclusion, which represents the source volume of an earthquake, where the rocks have fractured under increasing tectonic stresses, and thus has lost its shear strength. Since slow-slip events are ruptures, basically the same as earthquakes, the same model should be applicable to them also.

Water-level drops have been previously observed at some apparently sensitive wells before other earthquakes in Nankai trough. For example, 10 days before the 1946 magnitude 8 Nankai earthquake, groundwater in 11 out of more than 160 shallow wells along the Pacific coast from Kii Peninsula to Shikoku ran dry. Similar water-level decreases also occurred before the 1854 Nankai earthquake. These drops are also attributable to the occurrence of some pre-earthquake slow-slip events (Koizumi 2013 and references therein).

As to the 5 smaller pre-earthquake water-level drops (Fig. 6), they occurred at a time when there was no measurements made for slow-slip events in Japan. However, four of them did occur in the subduction zones that were later known to be producing such events. Two of them occurred before two earthquakes (M6.6 on September 24, 1990 and M6.1 on September 5, 1996) in Nankai trough and the other two before two earthquakes (M8.1 on October 4, 1994 and M7.5 on December 18, 1994) in northeastern Japan along the Japan Trench, where some slow-slip events have reportedly triggered some megathrust earthquakes including the 2011 magnitude 9 earthquake (Kato et al. 2012; Ito et al. 2013; Uchida et al. 2016). The calculated strain increments at Tono associated with three of these earthquakes are relative large, about $10^{-7}$ or more, as indicated by diamonds in Fig. 7. If their pre-slips in areas downdip the hypocenters (thus closer to Tono) had magnitudes of about 6 and 7 for Nankai trough and Japan Trench, respectively, as the slow-slip events observed later (e.g., Hirose and Obara 2005; Schwartz and Rokosky 2007; Kato et al. 2012; Ito et al. 2013; Uchida et al. 2016), then the estimated static strain increments they caused should be several tens of nanostrain (see Fig. 7), a level at which many earthquake-related hydrological changes were observed (threshold line in Fig. 7; Wang and Manga 2014). The remaining water-level drop in Fig. 6 occurred before the magnitude-7.2 Kobe earthquake on 1995/1/17 along an inland fault, rather than a subduction zone. For this earthquake also, there are reports of various anomalies indicating that some pre-earthquake slow-slip events may have occurred. For example, Igarashi et al. (1995) reported that, about 60 days before the earthquake groundwater radon concentration in a well located about 30 km away in the subsequent aftershock area began to rise to about 3 times above the background level of about 20 Bq/liter; about 10 days before the earthquake the concentration showed a wild drop-rise-drop change between the background level and 240 Bq/liter; and the concentration returned to background level after the earthquake. Tsunogai and Wakita (1995) analyzed groundwater samples collected on different dates from two 100-m deep wells in the Rokko mountain, and found chloride and sulfate concentrations in these samples to have begun to increase about 5 months before the earthquake to a peak value of about 10 standard deviations above the background level. They attributed these increases to the introduction of water from fracture zone to the source aquifer. Additionally, Yasuoka et al. (2006) monitored atmospheric-radon concentration at a site on the Rokko fault, and found an anomalous increase beginning several months before the earthquake. Both local and regional components of the concentration reached an anomalously high level about 17 days before the earthquake, then the local component decreased sharply while the regional component remained high until the occurrence of the earthquake; after the earthquake the radon concentration returned to normal. All these observations show that the local stress along the seismogenic fault may have reached the near-failure level, and caused some slow-slip events and foreshocks to occur along the shallower part of the fault about 2 weeks before the earthquake.

### 3.3. Different Permeability Changes

In the following, the pre-earthquake-, and leakage-induced water-level changes at SN3 are compared in detail to better understand the mechanisms responsible for these changes. As shown in Fig. 3, the coseismic water-level drops amounted to 10–50 cm; they occurred at a fast rate, beginning about 20 s after the respective earthquakes; they began to recover about 3 h later and lasted for 5–10 days.

For the magnitude 5.8 earthquake, the coseismic water-level drop of about 40 cm (Figs. 4, 5) also began about 20 s after the earthquake; it also began to recover 3 h later. But the recovery lasted longer (about 6 weeks) and reached a much higher level of 190 cm above that before the 6-month pre-seismic drop. This shows that the recovery probably consisted of two parts, one for the pre-seismic drop (about 150 cm) and the other for the coseismic drop (about 40 cm). The longer recovery time is probably due to the larger combined water-level drop to be recovered and the larger damage to the well-tapped aquifer/barrier system by the stronger seismic shaking, as discussed below.

The 5 smaller pre-earthquake drops (Fig. 6) amounted to 20–60 cm, beginning 5–15 days before the respective earthquakes, and their recoveries took 6–10 days. Some of the pre-earthquake drops (the upper two curve in Fig. 6) were superposed with coseismic drops of smaller amount, similar to the case of the magnitude-5.8 earthquake (Fig. 5). These recoveries did not always began after the respective earthquakes. Thus, the assumed pre-earthquake slow-slip events may have played a more dominating role than the earthquakes in causing these water-level changes.

The observation that the coseismic drops began sooner (20 s after the earthquakes) and proceeded at a much faster rate than the pre-earthquake drops is understandable, because the damage by seismic shaking occurred almost instantly, whereas those by slow-slip events occurred more gradually. Additionally, the seismic shaking may have opened up fissures not only in the fault zone but also in the aquitards of the well, and even damaged the annular zone around the well bore. Such a vertical breaching of aquitards was invoked by King et al. (1999) to explain the post-earthquake water-level rise at TH7 well, as mentioned above, and has been inferred for other wells also (see Wang et al. 2016). However, the coseismic recoveries took about the same amount of time to complete as in the pre-earthquake cases (5–10 days). This may be because the healings of the damaged aquitards and fault zone could proceed simultaneously. In contrast to their slower rate, the amount of

pre-earthquake water-level drops are slight larger than the coseismic drops. This may indicate a slightly larger spatial extent of damage to the fault zone caused by the slow-slip-induced stress increments than the seismic shakings.

In the leakage test, a water-level drop of about 150 cm at SN3 began about 9 h after the test began at the KNA2 hole. This 9-h delay suggests a permeability of $10^{-13}$ to $10^{-11}$ m$^2$ for the aquifer, if assuming a specific storage of $10^{-6}$ to $10^{-4}$ m$^{-1}$. The delay is much longer than the 20 s delay for the coseismic drops by seismic shaking, and the recovery rate was much smaller also. This again suggests a much larger-scaled and closer-to-the-well leakage caused by seismic shaking than the leakage at this hole. However, the 9-h delay is much shorter than the 5–15 days for the pre-earthquake drops, suggesting a slower damaging process for the pre-earthquake case. This indicates that the damage in this case was probably controlled mainly by the slow building-up of the stress increments by the slow-slip events. The recovery in the leakage test began about 1 day after the leak was stopped, and proceeded at a rate somewhat smaller than the pre-earthquake and coseismic cases. The 1-day delay is longer than the 3-h delay for the coseismic drops. This is probably because the effect of healing at locations closer to the well was sensed earlier in the coseismic case. The slightly smaller recovery rate in the leakage test probably indicates a smaller source area (a single hole) of the leak in the fault zone than the areas of multiple fissures in the fault zone in the coseismic and pre-seismic cases.

### 3.4. Long-Distance Triggering

Many of the earthquakes that showed coseismic water-level changes in this study are located far away from the monitored water wells, and this may raise question about their genuineness. However, long-distance triggering of various hydrological, geochemical and geophysical phenomena, such as earthquake, volcanic eruption, geyser eruption, and water-level and stream-flow changes, have been observed frequently now (e.g., Rojstaczer and Wolf 1992; King et al. 1995, 2006; Hill et al. 1993, 2002; Silver and Vallette-Silver 1992; Roeloffs 1996; Kitagawa and

Koizumi 1996; Gomberg et al. 2001; Segall and Bradley 2012; Wang and Manga 2014; Uchida et al. 2016; Obara and Kato 2016). The triggering agents include seismic shaking, stress increment generated by episodic slow slip, and even typhoon (Liu et al. 2009). The triggered changes were not necessarily observed at sites closest to the triggering event, suggesting also that they occurred only at sensitive sites that had already been in a critical condition. The existence of relatively few sensitive sites among many insensitive ones shows the inadequacy of poro-elasticity assumption for modeling the crust, and the importance of taking crustal inhomogeneity into consideration.

## 4. Concluding Remarks

The set of high-quality water-level data used here were continuously recoded at many closely clustered sites in a setting of favorable and well-known hydrogeological condition, including the existence of a fault with large pressure gradient, during a period of high seismic activity, when the rare magnitude 5.8 local earthquake and many larger more distant earthquakes occurred. It has thus provided a good opportunity for comparatively studying the coseismic, pre-earthquake, and post-earthquake changes in a heterogeneous crust in detail. However, at the time of data gathering of this study, there were not yet any significant geodetic and seismic monitoring measurements directed to studying slow-slip events at Tono, so that we may use to test the proposed mechanism. Yet, I hope the detailed description of the characteristics of the sensitive SN3 well presented here may demonstrate that genuine pre-earthquake hydrological changes can be observed at some "sensitive" sites under near-failure condition and that the underlying mechanisms can be understood.

Future research should include comparative studies of water-level and geodetic/seismic changes to better ascertain the effect of slow-slip events on both. In quantitative interpretation of water-level changes related to seismic and aseismic fault slips, it is important to keep in mind that the idealized poro-elastic model may not be realistic enough to represent the heterogeneous crust, thus its results should be used only as a guide rather than an absolute standard.

May the result reported here stimulate other efforts to search for, or even to create, similarly sensitive sites (such as, by drilling monitoring wells that tap aquifers closely connected to some critically stressed impermeable fault and by pumping out groundwater or injecting fluid on one side of the fault to create a large hydraulic-pressure gradient across it), so that more pre-earthquake changes may be observed and studied. With a concerted and well-planned long-term monitoring and research effort in some properly chosen seismic regions, it may be possible someday to show that prediction is possible for at least some earthquakes, especially those with pre-slips, when such slips are taken into consideration together with crustal dilatancy in the underlying mechanism (King and Chia 2017).

## Acknowledgements

I am thankful to my former colleagues at the University of Tokyo for providing the data used in this study, especially Y. Hitagawa for the Okada-model-based calculation of volumetric strain at Tono for the 1996–97 slow-slip event. I am also thankful to M. Manga and three anonymous reviewers for comments, which have helped to improve the presentation significantly.

## REFERENCES

Dobrovolsky, I. P., Zubkov, S. I., & Miacchkin, V. I. (1979). Estimation of the size of earthquake preparation zone. *Pure and Applied Geophysics, 117,* 1025–1044.

Gomberg, J., Reasonberg, P. A., Bodin, P., & Harris, R. A. (2001). Earthquake triggering by seismic waves following the Landers and Hector Mine earthquakes. *Nature, 411,* 462.

Hertzberg, R. W. (1989). *Deformation and fracture mechanics of engineering materials* (3rd ed., p. 680). New York: Wiley.

Hill, D. P., Pollitz, F., & Newhall, C. (2002). Earthquake-volcano interactions. *Physics Today, 55,* 41–47.

Hill, D. P., et al. (1993). Seismicity remotely triggered by the magnitude 7.3 Landers, California, earthquake. *Science, 260,* 1617–1623.

Hirose, H., & Obara, K. (2005). Repeating short- and long-term slow slip events with deep tremor activity around the Bungo channel region, southwest Japan. *Earth Planets Space, 57,* 961–972.

Igarashi, G., Saeki, S., Takahata, N., Sumikawa, K., Tasaka, S., Sasaki, Y., et al. (1995). Groundwater radon anomaly before the Kobe earthquake in Japan. *Science, 269,* 60–61.

Ito, Y., Hino, R., Fijimoto, H., Osada, Y., Inazu, D., Ohta, Y., et al. (2013). Episodic slow slip events in the Japan subduction zone before the 2011 Tohoku-Oki earthquake. *Tectonophysics, 600,* 14–26.

Kato, A., Obara, K., Igarashi, T., Tsuruoka, H., Nakagawa, S., & Hirata, N. (2012). Propagation of slow slip leading to the 2011 Mw 9.0 Tohoku-Oki earthquake. *Science, 335,* 705–708.

King, C.-Y. & Chia, Y. (2017). Anomalous streamflow and groundwater-level changes before the1999 M7.6 Chi-Chi earthquake in Taiwan: Possible mechanisms. *Pure and Applied Geophysics* https://doi.org/10.1007/s00024-017-1737-1.

King, C.-Y., Nason, R. D., & Tocher, D. (1973). Kinematics of fault creep. *Philosophical Transactions of the Royal Society London A., 274,* 355–360.

King, C.-Y., Koizumi, N., & Kitagawa, Y. (1995). Hydrogeochemical anomalies and the 1995 Kobe earthquake. *Science, 269,* 38–39.

King, C.-Y., Azuma, S., Igarashi, G., Ohno, M., Saito, H., & Wakita, H. (1999). Earthquake-related water-level changes at 16 closely clustered wells in Tono, central Japan. *Journal of Geophysical Research, 104*(B6), 13073–13082.

King, C.-Y., Azuma, S., Ohno, M., Asai, Y., He, P., Kitagawa, Y., et al. (2000). In search of earthquake precursors in the water-level data of 16 closely clustered wells at Tono, Japan. *Geophysical Journal International, 143,* 469–477.

King, C.-Y., Zhang, W., & Zhang, Z. (2006). Earthquake-induced groundwater and gas changes. *Pure and Applied Geophysics, 163,* 633–645.

Kitagawa, Y., & Koizumi, N. (1996). Comparison of post-seismic groundwater changes with earthquake-induced volumetric strain release: Yudani hot spring, Japan. *Geophysical Research Letters, 23,* 3147–3150.

Kobayashi, A. (2014). A long-term slow slip event from 1996 to 1997 in the Kii Channel, Japan, Earth, Planets. *Space, 66,* 9. https://doi.org/10.1186/1880-5981-66-9.

Koizumi, N. (2013). Earthquake prediction research based on observation of groundwater. *Synthesiology, 6,* 24–33. **(English translation)**.

Linde, A. T., Gladwin, M., Johnston, M., Gwyther, R., & Bilham, R. (1996). A slow earthquake sequence on the San Andreas Fault. *Nature, 383,* 65–68.

Linde, A. T., & Sacks, S. I. (2002). Slow earthquakes and great earthquakes along the Nankai Trough, Earth Planet Sci. *Lett., 203,* 265–275.

Linde, A. T., & Sacks, I. S. (2003). Slow earthquakes and great earthquakes along the Nankai trough, Earth and Planet. *Science Letters, 203,* 265–275.

Liu, C., Linde, A. T., & Sacks, S. (2009). Slow earthquakes triggered by typhoons. *Nature, 459,* 833–836.

Matsubara, M., Obara, K., & Kasahara, K. (2009). High-vp/vs zone accompanying non-volcanic tremors and slow-slip events beneath southwestern Japan. *Tectonophysics, 472,* 6–17.

Obara, K., & Kato, A. (2016). Connecting slow earthquakes to huge earthquakes. *Science, 353,* 253–257.

Okada, Y. (1992). Internal deformation due to shear and tensile faults in a half space. *Bulletin of the Seismological Society of America, 82*(2), 1018–1040.

Roeloffs, E. A. (1996). Poroelastic techniques in the study of earthquake-related hydrologic phenomena. *Advances in Geophysics, 37,* 135–195.

Rojstaczer, S., & Wolf, W. (1992). Permeability changes associated with large earthquakes: an example from Loma Prieta. *California Geology, 20,* 211–214.

Schwartz, S. Y., & Rokosky, J. M. (2007). Slow slip events and seismic tremors at circum-Pacific subduction zones. *Reviews Geophysics, 45,* RG3004. https://doi.org/10.1029/2006rg000208.

Segall, P., & Bradley, A. M. (2012). Slow-slip evolves into megathrust earthquakes in 2D numerical simulations. *Geophysical Research Letters, 39,* L18308.

Silver, P. G., & Vallette-Silver, N. J. (1992). Detection of hydrothermal precursors to large northern California earthquakes. *Science, 257,* 1363–1368.

Tsunogai, U., & Wakita, H. (1995). Precursory chemical changes in groundwater: Kobe earthquake, Japan. *Science, 269,* 61–63.

Uchida, N., Iinuma, T., Nadeau, R. M., Burgmann, R., & Hino, R. (2016). Periodic slow slip triggers megathrust zone earthquakes in northeastern Japan. *Science, 351,* 488–492.

Wakita, H. (1996). Geochemical challenge to earthquake prediction. *Proceedings of the National academy of Sciences of the United States of America, 93,* 3781–3786.

Wang, C., Liao, X., Wang, L.-P., Wang, C.-H., & Manga, M. (2016). Large earthquakes create vertical permeability by breaching aquitards. *Water Resources Res..* https://doi.org/10.1002/2016WR018893.

Wang, C., & Manga, M. (2014). Earthquakes and water. *Encyclopedia of Complex and Systems Science.* https://doi.org/10.1007/978-3-642-00810-8.

Yasuoka, Y., Igarashi, G., Ishikawa, T., Tokonami, S., & Shinogi, M. (2006). Evidence of precursor phenomena in the Kobe earthquake obtained from atmospheric radon concentration. *Applied Geochemistry, 21,* 1064–1072.

(Received July 25, 2017, revised March 26, 2018, accepted March 27, 2018)

Pure Appl. Geophys.
© 2017 The Author(s)
This article is an open access publication
DOI 10.1007/s00024-017-1670-3

| Pure and Applied Geophysics

# Streamflow Changes in the Vicinity of Seismogenic Fault After the 1999 Chi–Chi Earthquake

CHING-YI LIU,[1] YEEPING CHIA,[1] PO-YU CHUANG,[1,2] CHI-YUEN WANG,[3] SHEMIN GE,[4] and MAO-HUA TENG[1]

*Abstract*—Changes in streamflow have been observed at 23 stream gauges in central Taiwan after the 1999 $M_W$ 7.6 Chi–Chi earthquake. Post-earthquake increases, ranging from 58 to 833% in discharge, were recorded at 22 gauges on four rivers and their tributaries. The streamflow increase typically peaked in 2–3 days and followed by a slow decay for a month or more. An increased groundwater discharge to the river after the earthquake can be attributed to rock fracturing by seismic shaking as well as pore pressure rise due to compressive strain. A large decrease in discharge was recorded immediately after the earthquake at the gauge near the earthquake epicenter. Further analysis of long-term data indicates that the post-earthquake discharge at the gauge reduced to a level smaller than that at an upstream gauge for 8 months. Such a streamflow decrease might have been caused by a discharge to the streambed due to a co-seismic decrease in pore pressure induced by crustal extension during the rupture of the thrust fault.

**Key words:** Streamflow, gauge, Chi–Chi earthquake, deformation, groundwater.

## 1. Introduction

Hydrological changes associated with earthquakes have been observed at many places in the world. Generally, changes in groundwater level occur co-seismically (Waller 1966; Roeloffs 1998; King and Igarashi 1999; Chia et al. 2001; Cox et al. 2012; Weingarten and Ge 2014, He et al. 2017), while changes in streamflow or spring flow appear after major earthquakes (Rojstaczer and Wolf 1992; Muir-Wood and King 1993; King et al. 1994; Manga 2001; Manga et al. 2003; Montgomery et al. 2003; Wang et al. 2004a, b; Manga and Rowland 2009; Mohr et al. 2012, 2017).

Earthquake-triggered hydrological changes may provide insight into the mechanism underlying the crustal processes during the fault rupture. Groundwater-level changes in the vicinity of seismogenic fault are generally attributed to the redistribution of stress–strain due to fault movement (Roeloffs 1988; Montgomery and Manga 2003). Rojstaczer and Wolf (1992) proposed that streamflow increases are likely caused by the increase in permeability of near-surface rock due to fracturing by seismic shaking. Muir-Wood and King (1993) related the hydrological changes following major earthquakes to the style of faulting. Significant streamflow increases were found to accompany major normal fault earthquakes, while small decreases were associated with reverse fault earthquakes. Soil liquefaction has been proposed for the streamflow increase (Manga 2001; Wang et al. 2001; Montgomery et al. 2003), but possibly limited to shallow sediments in the river valley of the mountain area (Mohr et al. 2012).

During the 1999 $M_W$ 7.6 Chi–Chi earthquake, the largest and most destructive inland earthquake in the history of Taiwan, stream discharges recorded by a dense network of gauges on major rivers and their tributaries provide comprehensive data for exploring earthquake-triggered hydrological changes. In this paper, we presented the streamflow changes in the vicinity of the seismogenic fault before and after the Chi–Chi earthquake. Long-term data analysis and further investigations provide a basis for a better understanding of the natural processes of the earthquake hydrological anomalies.

[1] Department of Geosciences, National Taiwan University, No. 1 Section 4 Roosevelt Road, Taipei 106, Taiwan. E-mail: ypc@ntu.edu.tw
[2] Geotechnical Engineering Research Center, Sinotech Engineering Consultants, Inc, Taipei 114, Taiwan.
[3] Department of Earth and Planetary Science, University of California, Berkeley, CA 94720, USA.
[4] Department of Geological Sciences, University of Colorado, Boulder, CO 80309, USA.

## 2. Seismicity, Geology and Hydrology

As a part of the Circum-Pacific seismic belt, Taiwan is located at the convergent boundary between the Philippine Sea plate and the Eurasian plate. As a result of movement and collisions of the two plates, Taiwan is one of the most seismically active regions in the world. Historical destructive earthquakes could be documented back to 1644, while earthquakes recorded by seismic instruments started after 1897 (Hsu 1980). From 1990 to 2006, there were 899 earthquakes cataloged with magnitude greater than $M_W$ 5.5 (Chen and Tsai 2008).

The island of Taiwan is located at the margin of young orogenic belt containing a series of north-northeast trend geologic formations (Ho 1986). The western plain consists of a thick sequence of undeformed unconsolidated deposits, while sedimentary and metamorphic rocks appear along with complex fold and fault structures in the mountainous area. As mountain ranges in the island stretch from north to south, most rivers flow in the east–west direction. Groundwater storage is different in the consolidated formation and the unconsolidated sediments (WRA 2016). In the mountains area, groundwater is stored in the pores of sedimentary rocks as well as fractures, such as joints and faults, of the metamorphic rocks. Major aquifers, however, are located in gravel and sand layers of unconsolidated deposits in the western plain.

## 3. Chi–Chi Earthquake and Co-seismic Groundwater-Level Changes

The 1999 Chi–Chi earthquake occurred at 1:47 am on September 21, 1999, local time (17:47 on 20 September UTC). The epicenter of the $M_W$ 7.6 earthquake is located at 23.85°N; 120.82°E, near the town of Chi–Chi in central Taiwan (Fig. 1). The depth of the hypocenter of the earthquake is about 8 km. The focal mechanism of the mainshock was of a thrust type with strike 5°, dip 34°. Surface rupture appeared to extend approximately 100 km in the north–south direction along the traces of the Chelungpu thrust fault (Angeliera et al. 2003). Horizontal and vertical offset caused by crustal deformation

ranged from 2.4 to 10.1 m and 1.2 to 4.4 m across the Chelungpu fault, respectively. In the epicentral region, the crustal deformation was essentially a uniaxial compressional strain of 0.36 micro-strain/year in the direction of 114° over several years before the earthquake (Yu et al. 2001). After the mainshock, more than 10,000 aftershocks were recorded in the first 3 weeks (Ma et al. 1999).

In the previous studies, examination of available data recorded at monitoring wells in the footwall of the Chelungpu fault revealed widespread co-seismic groundwater-level changes (Wang et al. 2001). Widespread sustained co-seismic groundwater-level changes were observed in monitoring wells in the footwall (Chia et al. 2001, 2008a; Wang et al. 2001). As shown in Fig. 1, groundwater-level falls were recorded in wells near the ruptured fault, while rises were recorded in most wells farther away to the west in the coastal plain consisting of unconsolidated deposits (Chia et al. 2001). Generally, larger changes, up to 11.09 m, were found in wells closer to the fault. These co-seismic changes were likely caused by co-seismic strain induced by fault displacement (Wakita 1975; Roeloffs 1988; Grecksch et al. 1999; Ge and Stover 2000; Chia et al. 2001, 2008b). The distribution of the sustained changes in the footwall revealed that crustal extension dominated near the Chelungpu fault during the earthquake, while compression prevailed away from the fault.

## 4. Rainfall and Stream Gauges

In central Taiwan, the average annual precipitation is about 2154 mm, but it may reach to 4000 mm in the mountainous area. The climate of the area is characterized by a distinct wet and dry season, with most of the rainfall occurring between May and September. Heavy rainfall usually occurs in short duration, and thus, most stream discharges peak shortly after rainfall. The river usually takes a few days to return to the base flow. Significant rainfall-induced changes in streamflow are frequently recorded during wet seasons. The Chi–Chi earthquake occurred at the end of the wet season. During the dry season, the gradual decay in discharge reflects diminishing supply from the groundwater discharge.

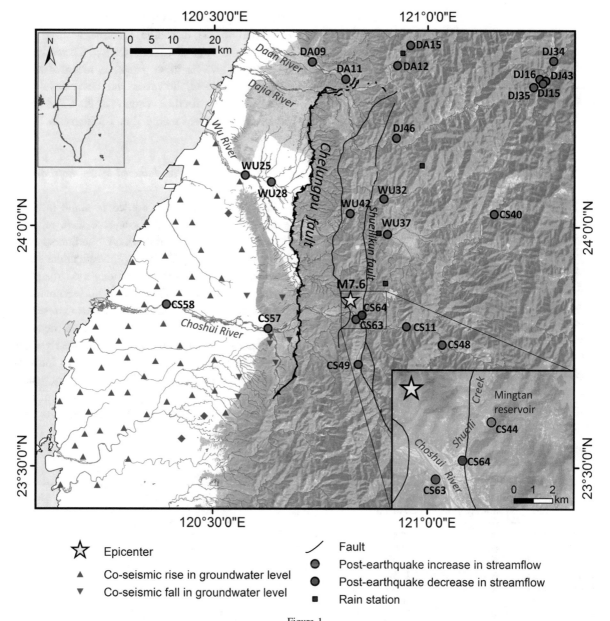

Figure 1
Distribution of hydrological monitoring stations during the 1999 $M_W$ 7.6 Chi–Chi earthquake in central Taiwan. The upper left inset shows the area in Taiwan. Post-earthquake streamflow increases (blue circles) and decrease (red circle) were observed at 23 stream gauges. Co-seismic groundwater-level rises (blue triangles) and falls (orange inverted triangles) were observed in 152 wells. The inset map in the lower right shows the location of gauge stations CS64, CS44, CS63 and the earthquake epicenter

During the Chi–Chi earthquake, streamflow discharges were recorded at 32 gauges on four major rivers, including Daan River, Dajia River, Wu River and Choshui River, and their tributaries in the vicinity of the Chelungpu fault (Fig. 1). The four rivers originated from the Central Mountain Ranges, flowing westward through the coastal plain to the Taiwan Strait. The length of these rivers ranges from 96 to 187 km, and the watershed area ranges from 758 to 3157 km² (Table 1). Of those, the Choshui River is the longest river in Taiwan.

Table 1

*Stream gauges on four rivers in central Taiwan*

| River | Length (km) | Drainage area (km$^2$) | Gauge name |
|---|---|---|---|
| Daan | 95.7 | 758.4 | DA09, DA12, DA11, DA15 |
| Dajia | 124.2 | 1235.7 | DJ08, DJ14, DJ15, DJ16, DJ34, DJ35, DJ37, DJ43, DJ45, DJ46 |
| Wu | 119.1 | 2025.6 | WU25, WU28, WU32, WU37, WU38, WU42 |
| Choshui | 186.6 | 3156.9 | CS07, CS11, CS24, CS40, CS44, CS48, CS49, CS57, CS58, CS61, CS63, CS64 |

These gauges were installed by the Water Resource Agency for the water resource management and flood prevention. Most of these gauges are located on the mountainous east, or hanging wall, side of the fault. They were used to measure the elevation of the stream water surface. Water level sensor system was designed to fit the characteristics of the cross section of the river at these gauges. A water-stage recorder was installed on or near the bridge crossing the river for automatically recording water level of the stream. The water level was then converted to the discharge (volumetric flow rate) based on the stage-discharge relation. Currently, the river water level at these gauges is recorded at 1-h intervals. In 1999, however, the water level was recorded daily, with the exception of the CS44 gauge where the data were recorded at 1-h intervals.

## 5. Streamflow Changes Induced by the Earthquake

Streamflow changes recorded at some of these gauges after the Chi–Chi earthquake were reported before (Wang et al. 2004b). As additional data became available and further investigations were conducted, it was found that, among the streamflow changes recorded at 32 changes, six anomalously large increases were caused by upstream reservoir release for the dam safety, one substantial decrease was caused by a large landslide dam on the river due to the earthquake, and two changes were indistinguishable from rainfall disturbances or daily fluctuation. Thus, only 23 gauges recorded distinct earthquake-related changes. Of those, 22 are post-

Table 2

*Post-earthquake streamflow increase at 22 stream gauges*

| Gauge name | Elevation (m) | Epicentral distance (km) | Discharge on September 20 (m$^3$/s) | Peak discharge (m$^3$/s) | Days to reach peak discharge | Percentage change in discharge (%) |
|---|---|---|---|---|---|---|
| DJ15 | 1444 | 64 | 9.4 | 15.2 | 2 | 62 |
| DJ16 | 1451 | 64 | 5.8 | 12.4 | 3 | 114 |
| DJ34 | 1629 | 69 | 4.5 | 7.1 | 3 | 58 |
| DJ35 | 1434 | 62 | 16.5 | 30.2 | 3 | 83 |
| DJ43 | 1472 | 65 | 3.5 | 6.9 | 3 | 97 |
| DJ46 | 550 | 35 | 0.3 | 2.8 | 3 | 833 |
| WU25 | 10 | 36 | 103.0 | 168.0 | 3 | 63 |
| WU28 | 25 | 31 | 9.6 | 49.2 | 2 | 413 |
| WU32 | 334 | 22 | 16.4 | 84.2 | 3 | 413 |
| WU37 | 379 | 16 | 28.4 | 68.6 | 2 | 142 |
| WU42 | 220 | 18 | 43.0 | 119.0 | 2 | 177 |
| DA09 | 189 | 50 | 14.2 | 34.0 | 1 | 139 |
| DA11 | 325 | 46 | 15.3 | 36.6 | 2 | 139 |
| DA12 | 598 | 50 | 2.1 | 6.6 | 4 | 214 |
| DA15 | 531 | 55 | 8.3 | 34.1 | 2 | 311 |
| CS11 | 364 | 14 | 82.9 | 133.8 | 3 | 61 |
| CS40 | 1055 | 38 | 4.7 | 10.3 | 3 | 119 |
| CS48 | 302 | 7 | 26.4 | 42.4 | 2 | 61 |
| CS49 | 475 | 14 | 17.5 | 36.8 | 4 | 110 |
| CS57 | 106 | 20 | 128.0 | 248.0 | 2 | 94 |
| CS58 | 28 | 43 | 152.0 | 304.0 | 3 | 100 |
| CS63 | 278 | 4 | 86.0 | 209.0 | 2 | 143 |

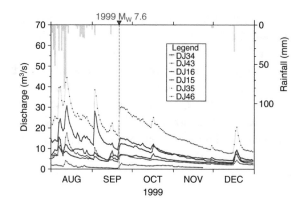

Figure 2

Daily discharge from August to December 1999 at six gauges on the Dajia River and its tributaries. Streamflow increases, ranging from 58 to 833%, were recorded after the earthquake

earthquake streamflow increases and the remaining one is post-earthquake decrease (Table 2). The distance between these gauges and the epicenter ranges from 4 to 69 km.

### 5.1. Post-earthquake Streamflow Increase

Six gauges on the Dajia River and its tributaries recorded a streamflow increase, ranging from 58 to 833%, in 2 to 3 days after the earthquake (Table 2). The rapid increase was followed by a slow decay over the next month or more. The temporal variations of daily discharge at these gauges are shown in hydrographs, along with nearby rainfall data (Fig. 2). The largest percentage increase was recorded at DJ46, with an elevation of 550 m, located on a

tributary in the downstream of the Dajia River. The discharge increased 833% to 2.8 $m^3/s$ 3 days after the earthquake. The large percentage increase was primarily caused by the relatively small discharge, approximately 0.3 $m^3/s$, on the day before the earthquake. The other five gauges are located near the source of the Dajia River with an elevation between 1434 and 1629 m. For instance, DJ35 is located on the mainstream for monitoring the sum of upstream tributary discharges. An increase of 83% in discharge, from 16.5 $m^3/s$ on the day before the earthquake to a peak flow of 30.2 $m^3/s$ 3 days after the earthquake, was recorded at the gauge (Fig. 2). The peak discharge took more than 1 month to return to the pre-earthquake value. The slow decay could be disturbed by rainfall, such as an increase in discharge following the heavy rainfall in mid-December.

The temporal variations of daily discharge at five gauges on the Wu River and its tributaries show streamflow increases, ranging from 63 to 413%, after the earthquake (Fig. 3). At WU28, an increase of 413% in discharge was recorded on the mainstream. The discharge increased from 9.6 $m^3/s$ on the day before the earthquake to a peak value of 49.2 $m^3/s$ 2 days after the earthquake. The WU25 gauge, with an elevation of 10 m, is located on the mainstream in the coastal plain. It recorded a streamflow increase of 63%, from 103.0 $m^3/s$ to 168.0 $m^3/s$, in 3 days after the earthquake.

Four gauges on the Daan River and its tributaries recorded a streamflow increase, ranging from 139 to

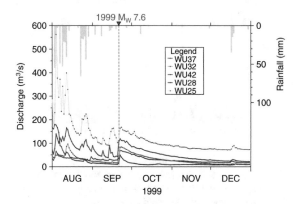

Figure 3

Daily discharge from August to December 1999 at five gauges on the Wu River and its tributaries. Streamflow increases, ranging from 63 to 413%, were recorded after the earthquake

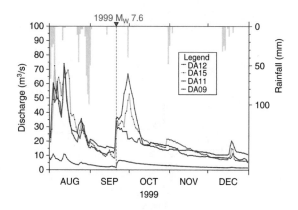

Figure 4

Daily discharge from August to December 1999 at four gauges on the Daan River and its tributaries. Streamflow increases, ranging from 139 to 311%, were recorded after the earthquake

311%, 1 to 4 days without rainfall after the earthquake (Table 2). The temporal variations of daily discharge at these gauges are shown in Fig. 4. At gauge DA12, for instance, an increase of 214% in discharge was recorded on a small tributary. The discharge increased from 2.1 m³/s on the day before the earthquake to a peak value of 6.6 m³/s 4 days afterward. The peak flow took about 2 months to return to the pre-earthquake level. At DA15, the discharge increased to the first peak value of 34.1 m³/s 2 days after the earthquake and then decayed during the third day. Instead of a gradual decay, the discharge increased again for a week to the second peak. Further investigations indicated that several small landslide dams and lakes were created in the upstream of DA15 during the earthquake (Lin et al. 2000; Chen and Shi 2000). The collapse of these dams and the rainfall at end of September provided the discharge for the second increase to the peak of 52.8 m³/s on October 1. A similar pattern of post-earthquake streamflow changes was observed at DA11, which is located in the downstream of DA15.

Seven gauges on Choshui River and its tributaries recorded streamflow increases after the earthquake (Fig. 5). Of those, four gauges, CS11, CS63, CS57 and CS58, were installed on the mainstream, while the other three, CS40, CS48 and CS49, were on the tributaries where the discharge is relatively smaller. The streamflow increases at these gauges, ranging from 61 to 143%, are listed in Table 2. For instance,

at CS48, the discharge increased from 26.4 m³/s on the day before the earthquake to a peak value of 42.4 m³/s, or an increase of 61%, over 2 days after the earthquake. The increased streamflow took about one and a half month to return to the pre-earthquake level. At CS11 and CS49, however, two post-earthquake peak flows were recorded after the earthquake. Further investigations indicated the anomalous changes were also caused by the collapse of landslide dams and local rainfall in the upstream.

## 5.2. Post-earthquake Streamflow Decrease

After the 1999 Chi–Chi earthquake, the only streamflow decrease was observed at gauge CS64. This gauge, located only 4 km to the southeast of the epicenter, was placed to monitor streamflow in the downstream of the Shueili Creek (inset of Fig. 1). The Shueili Creek, a tributary of the Choshui River, lies almost directly above the hypocentral area and coincides approximately with the Shueilikun fault. Its bedrock is composed of fractured quartzitic sandstone and arkosic sandstone interbedded with argillite. The discharge recorded at CS64, as shown in Fig. 6a,

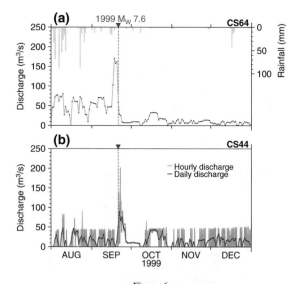

Figure 6
Time series of daily discharges recorded between August and December 1999 at CS64 and CS44. **a** The daily hydrograph at CS64 (orange line) showing the discharge decreased 96% immediately after the earthquake. **b** The daily (blue line) and hourly (gray line) hydrograph at CS44 showing the discharge was primarily affected by occasional water releases from the nearby Mingtan reservoir on the Shueili Creek

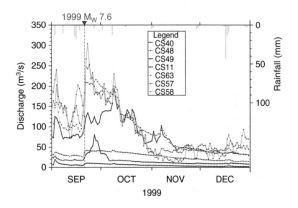

Figure 5
Daily discharge from September to December 1999 at seven gauges on the Choshui River and its tributaries. Streamflow increases, ranging from 61 to 143%, were recorded after the earthquake

displayed a large and abrupt decrease, from 162.0 to 58.8 m$^3$/s, on the day of the Chi–Chi earthquake. During the following few days, it further decreased to 7.1 m$^3$/s.

There was another gauge, CS44, located near the base of the Mingtan Dam approximately 3-km upstream from CS64 along the Shueili Creek (inset of Fig. 1). CS44 was installed by the Taiwan Power Company to monitor the release of Mingtan reservoir to the Shueili Creek. The hourly discharge at CS44, as shown in Fig. 6b, was normally maintained below 50 m$^3$/s, except for emergency need or during heavy rainfall. It did not show either an abrupt post-earthquake decrease, as recorded by CS64. In fact, the daily discharge at CS44 increased from 14.1 m$^3$/s immediately before the earthquake to 89.6 m$^3$/s after the earthquake. The hourly discharge also showed a rapid increase, up to 202.5 m$^3$/s, after the earthquake (Fig. 6b). The increase in discharge or reservoir release was imposed for dam safety. However, such a large increase at CS44 did not cause an increase in discharge at the downstream gauge CS64. Instead, the post-earthquake discharge continued to decline.

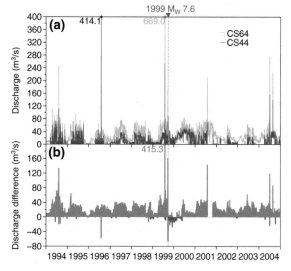

Figure 7
Daily discharges at two gauges, CS64 and CS44, on the Shueili Creek from 1994 to 2004. **a** Downstream discharges at CS64 (orange line) and upstream discharges at CS44 (blue line). **b** The difference in discharges between CS64 and CS44. The daily discharge at CS64 is normally larger than that at CS44 upstream (green color), but the situation was reversed after the Chi–Chi earthquake for 8 months (red color), indicating creek water loss to the ground between CS44 and CS64

The long-term data provide a more detailed comparison in discharge between the gauges CS64 and CS44, as shown in Fig. 7a. Generally, the discharge at the two gauges shows a similar seasonal pattern that increases in the wet seasons and decreases in the dry seasons. It is also noted that, similar to most rivers that obtain their water from the groundwater discharge during the dry seasons, the discharge of the Shueili Creek at the downstream gauge CS64 was usually larger than that at the upstream gauge CS44. However, a reversed situation appeared between September 1999 and June 2000 (Fig. 7b). The reversal event suggests that the Shueili Creek was changed to a influent (losing) stream from a effluent (gaining) stream immediately after the occurrence of the earthquake on September 21, 1999. Such a phenomenon lasts until the wet season arrived in May 2000. Therefore, based on the long-term monitoring data, the Shueili Creek had lost water to the creek bed over the 3-km segment between the gauges CS44 and CS64 in the 8 months after the earthquake.

Near CS64 was the other gauge, CS63, which recorded no streamflow decrease, but a post-earthquake increase, like most other gauges (Fig. 5). This is not surprising, because CS63 is located on the mainstream of the Choshui River, the recorded data represent the sum of discharges from all upstream tributaries (inset of Fig. 1).

## 6. Discussions

The post-earthquake streamflow increases recorded by the 22 gauges are similar to those observed in the previous studies. The only source of the increases with no rainfall comes from the discharge of groundwater. Based on Darcy's law, the increase of groundwater discharge to the river is caused by either an increase of rock permeability in the hills or a rise of pore pressure in the ground. The sudden increase of permeability is often explained by rock fracturing due to seismic shaking (Rojstaczer and Wolf 1992; Rojstaczer et al. 1995; Sato et al. 2000; Wang et al. 2004a, b; Charmoille et al. 2005; Elkhoury et al. 2006; Wang and Manga 2010; Manga et al. 2012; Mohr et al. 2015). Such an explanation can be supported by the occurrence of numerous landslides in the mountain area during the 1999 Chi–Chi earthquake (Hung 2000). For instance, the removal of vegetation and subsoil took place at 99 peaks during the earthquake, exposed the bedrock to the surface and enabled the groundwater to move through fractures more easily. Part of the streamflow increase may also be attributed to a rise of pore pressure induced by compressive strain during the fault rupture. Such an increase can be supported by the co-seismic groundwater-level rise prevailed in the footwall or the plain area. The mountain ranges in Taiwan were formed primarily by the compression between the Eurasian plate and the Philippine Sea plate. However, groundwater-level data were not recorded in the mountainous hanging wall during the Chi–Chi earthquake to support the compressive strain.

The earthquake-triggered streamflow increase is similar to the rainfall-induced increase, typically peaked in a few days. However, unlike a rapid decrease in discharge following the peak flow after rainfall, the post-earthquake peak flow is followed by a slow decay which may last for several months (Figs. 2, 3, 4). The slow decay is attributed to a prolonged groundwater discharge increase, instead of runoff, to the river. Such a groundwater discharge increase is consistent with an increase in either permeability or pore pressure proposed for the post-earthquake streamflow increase.

A large streamflow decrease triggered by the earthquake is rarely observed. Muir-Wood and King (1993) suggested that hydrological changes accompanying thrust fault earthquakes are most notable by their absence, or a decrease of well water level and spring flow. The field observation in the vicinity of the Chelungpu thrust fault, however, indicated post-earthquake streamflow increases at most gauges in the hanging wall. The only post-earthquake decrease in streamflow was recorded at the gauge nearest to the epicenter. Moreover, co-seismic falls in groundwater level were found primarily at wells adjacent to the thrust fault. Therefore, the crustal extension was likely to dominate from the epicentral region to the area adjacent to the thrust fault during the earthquake.

The long-term data indicate that the discharge at CS64 in the downstream of the Shueili Creek is usually larger than that at CS44 in the upstream. After the earthquake, however, the discharge at CS64 reduced to a level smaller than that at CS44 for 8 months or more. Such a post-seismic streamflow decrease might have been caused by a co-seismic decrease in pore pressure induced by crustal extension near the earthquake epicenter during the rupture of the thrust fault. The sudden decrease created a rapid downward discharge to the crust through the opening of pre-existing fractures in the bedrock of the Shueili Creek valley and new fractures generated by the earthquake. Such a downward discharge due to crustal extension must be far greater than the recharge from nearby hills due to seismic shaking. The sustained downward flow from surface water to the crust along the creek is likely to result in a streamflow in the downstream smaller than that in the upstream. The crustal extension is likely to extend to the area adjacent to the thrust fault, where co-seismic groundwater-level falls were recorded.

## 7. Conclusions

The significant streamflow decrease at CS64 on the Shueili Creek after the Chi–Chi earthquake is a unique observation. The gauge is only 4 km from the earthquake epicenter, and the Shueili Creek lies just above the hypocentral area. The proximity of gauge CS64 to the epicenter suggests that the streamflow decrease is likely induced by the crustal extension in the hanging wall during the rupture of the thrust fault. In the previous study (Chia et al. 2001), the co-seismic falls in groundwater level in the footwall near the Chelungpu fault trace were also attributed to the extensional deformation. Apparently, these hydrological anomalies reveal the dominance of the crustal extension in the area adjacent to the thrust fault in both the hanging wall and the footwall during the earthquake.

Beyond the zone of extension in central Taiwan, post-earthquake streamflow increase was recorded at all stream gauges and co-seismic groundwater-level rise was observed in nearly all monitoring wells. While the groundwater-level rise in the footwall was attributed to the compressive deformation, the streamflow increase in the hanging wall is possibly caused by both the permeability increase due to rock fracturing by seismic shaking and the pore pressure rise due to compressive deformation induced by fault displacement.

## Acknowledgements

The authors would like to thank the reviewers for the helpful comments and suggestions. We gratefully acknowledge access to stream discharge data of the Water Resources Agency of Taiwan and the Taiwan Power Company. This work is supported by the Ministry of Science and Technology of Taiwan (MOST 105-2116-M-002-023). Special thanks are extended to the members of the hydrogeology laboratory of the National Taiwan University for their research support.

## REFERENCES

Angeliera, J., Lee, J. C., Chu, H. T., & Hu, J. C. (2003). Reconstruction of fault slip of the September 21st, 1999, Taiwan earthquake in the asphalted surface of a car park, and co-seismic slip partitioning. *Journal of Structural Geology, 25*(3), 345–350.

Charmoille, A., Fabbri, O., Mudry, J., Guglielmi, Y., & Bertrand, C. (2005). Post-seismic permeability change in a shallow fractured aquifer following a $M_L$ 5.1 earthquake (Fourbanne karst aquifer, Jura outermost thrust unit, eastern France). *Geophysical Research Letters, 32*(18), L18406.

Chen, S. C., & Shi, D. R. (2000). *A documentary of soil and water hazard and rehabilitation of slope failure after the Chi-Chi earthquake*. Nantou City: Soil and Water Conservation Bureau. **(in Chinese)**.

Chen, K. P., & Tsai, Y. B. (2008). A Catalog of Taiwan Earthquakes (1900–2006) with Homogenized $M_W$ Magnitudes. *Bulletin of the Seismological Society of America, 98*(1), 483–489.

Chia, Y., Chiu, J. J., Chiang, Y. H., Lee, T. P., & Liu, C. W. (2008a). Spatial and temporal changes of groundwater level induced by thrust faulting. *Pure and Applied Geophysics, 165*(1), 5–16.

Chia, Y., Chiu, J. J., Chiang, Y. H., Lee, T. P., Wu, Y. M., & Horng, M. J. (2008b). Implications of coseismic groundwater level changes observed at multiple-well monitoring stations. *Geophysical Journal International, 172*(1), 293–301.

Chia, Y., Wang, Y. S., Chiu, J. J., & Liu, C. W. (2001). Changes of groundwater level due to the 1999 Chi–Chi earthquake in the Choshui River alluvial fan in Taiwan. *Bulletin of the Seismological Society of America, 91*(5), 1062–1068.

Cox, S. C., Rutter, H. K., Sims, A., Manga, M., Weir, J. J., Ezzy, T., et al. (2012). Hydrological effects of the $M_W$ 7.1 Darfield (Canterbury) earthquake, 4 September 2010, New Zealand. *New Zealand Journal of Geology and Geophysics, 55*(3), 231–247.

Elkhoury, J. E., Brodsky, E. E., & Agnew, D. C. (2006). Seismic waves increase permeability. *Nature, 441,* 1135–1138.

Ge, S., & Stover, S. C. (2000). Hydrodynamic response to strike- and dip-slip faulting in a half space. *Journal of Geophysical Research, 105*(B11), 25513–25524.

Grecksch, G., Roth, F., & Kümpel, H. J. (1999). Co-seismic well-level changes due to the 1992 Roermond earthquake compared to static deformation of half-space solutions. *Geophysical Journal International, 138*(2), 470–478.

He, A., Zhao, G., Sun, Z., & Singh, R. P. (2017). Co-seismic multilayer water temperature and water level changes associated with Wenchuan and Tohoku-Oki earthquakes in the Chuan no. 03 well, China. *Journal of Seismology, 21*(4), 719–734.

Ho, C. S. (1986). A synthesis of the geologic evolution of Taiwan. *Tectonophysics, 125*(1–3), 1–16.

Hsu, M. T. (1980). Destructive earthquakes in Taiwan—from 1644 to the present time. *Meteorological Bulletin, 26*(3), 32–48. **(in Chinese)**.

Hung, J. J. (2000). Chi–Chi earthquake induced landslides in Taiwan. *Earthquake Engineering and Engineering Seismology, 2*(2), 25–33.

King, C.-Y., Basler, D., Presser, T. S., Evans, W. C., & White, L. D. (1994). In search of earthquake-related hydrologic and chemical changes along Hayward fault. *Applied Geochemistry, 9*(1), 83–91.

King, C.-Y., & Igarashi, G. (1999). Earthquake-related water-level changes at 16 closely clustered wells in Tono, central Japan. *Journal of Geophysical Research, 104*(B6), 13073–13082.

Lin, M. L., Yu, F. C., Fan, J. C., & Lin, P. S. (2000). *Preliminary evaluation and analysis of potential debris flow after the Chi-Chi earthquake hazard*. Taipei: National Center for Research on Earthquake Engineering. **(in Chinese)**.

Ma, K. F., Lee, C. T., Tsai, Y. B., Shin, T. C., & Mori, J. (1999). The Chi–Chi, Taiwan earthquake: large surface displacements on an inland thrust fault. *EOS, 80*(50), 605–620.

Manga, M. (2001). Origin of postseismic streamflow changes inferred from baseflow recession and magnitude-distance relations. *Journal of Geophysical Research, 28*(10), 2133–2136.

Manga, M., Beresnev, I., Brodsky, E. E., Elkhoury, J. E., Elsworth, D., Ingebritsen, S. E., et al. (2012). Changes in permeability caused by transient stresses: field observations experiments and mechanisms. *Reviews of Geophysics, 50*(2), 2004.

Manga, M., Brodsky, E. E., & Boone, M. (2003). Response of streamflow to multiple earthquakes. *Geophysical Research Letters, 30*(5), 1214.

Manga, M., & Rowland, J. C. (2009). Response of Alum Rock springs to the October 30, 2007 Alum Rock earthquake and implications for the origin of increased discharge after earthquakes. *Geofluids, 9*(3), 237–250.

Mohr, C. H., Manga, M., Wang, C. Y., Kirchner, J. W., & Bronstert, A. (2015). Shaking water out of soil. *Geology, 43*(3), 207–210.

Mohr, C. H., Manga, M., Wang, C. Y., & Korup, O. (2017). Regional changes in streamflow after a megathrust earthquake. *Earth and Planetary Science Letters, 458,* 418–428.

Mohr, C. H., Montgomery, D. R., Huber, A., Bronstert, A., & Iroumé, A. (2012). Streamflow response in small upland catchments in the Chilean coastal range to the $M_W$ 8.8 Maule earthquake on 27 February 2010. *Journal of Geophysical Research, 117*(F2), F02032.

Montgomery, D. R., Greenberg, H. M., & Smith, D. T. (2003). Streamflow response to the Nisqually earthquake. *Earth and Planetary Science Letters, 209*(1–2), 19–28.

Montgomery, D. R., & Manga, M. (2003). Streamflow and water well responses to earthquakes. *Science, 300*(5628), 2047–2049.

Muir-Wood, R., & King, G. C. P. (1993). Hydrological signatures of earthquake strain. *Journal of Geophysical Research, 98*(B12), 22035–22068.

Roeloffs, E. A. (1988). Hydrologic precursors to earthquakes: a review. *Pure and Applied Geophysics, 126*(2), 177–209.

Roeloffs, E. A. (1998). Persistent water level changes in a well near Parkfield, California, due to local and distant earthquakes. *Journal of Geophysical Research, 103*(B1), 869–889.

Rojstaczer, S., & Wolf, S. (1992). Permeability changes associated with large earthquakes: an example from Loma Prieta. *California Geology, 20*(3), 211–214.

Rojstaczer, S., Wolf, S., & Michel, R. (1995). Permeability enhancement in the shallow crust as a cause of earthquake-induced hydrologic changes. *Nature, 373,* 237–239.

Sato, T., Sakai, R., Furuya, K., & Kodama, T. (2000). Co-seismic spring flow changes associated with the 1995 Kobe earthquake. *Geophysical Research Letters, 27*(8), 1219–1222.

Wakita, H. (1975). Water wells as possible indicators of tectonic strain. *Science, 189*(4202), 553–555.

Waller, R. M. (1966). *Effects of the March 1964 Alaska earthquake on the hydrology of Anchorage area.* USGS Prof. Pap. 544B

Wang, C. Y., Cheng, L. H., Chin, C. V., & Yu, S. B. (2001). Coseismic hydrologic response of an alluvial fan to the 1999 Chi–Chi earthquake, Taiwan. *Geology, 29*(9), 831–834.

Wang, C. Y., & Manga, M. (2010). Hydrologic responses to earthquakes and a general metric. *Geofluids, 10*(1–2), 206–216.

Wang, C. Y., Manga, M., Dreger, D., & Wong, A. (2004a). Streamflow increase due to rupturing of hydrothermal reservoirs: evidence from the 2003 San Simeon, California, earthquake. *Geophysical Research Letters, 31*(10), L10502.

Wang, C. Y., Wang, C. H., & Manga, M. (2004b). Co-seismic release of water from mountains: evidence from the 1999 ($M_w$ = 7.5) Chi–Chi, Taiwan, earthquake. *Geology, 32*(9), 769–772.

Water Resources Agency. (2016). *Hydrological yearbook of Taiwan Republic of China 2015 Total Report.* Taipei: Water Resources Agency, Ministry of Economic Affairs. **(in Chinese)**.

Weingarten, M., & Ge, S. (2014). Insights into water level response to seismic waves: a 24 year high-fidelity record of global seismicity at Devils Hole. *Geophysical Research Letters, 41*(1), 74–80.

Yu, S. B., Kuo, L. C., Hsu, Y. J., Su, H. H., Liu, C. C., Hou, C. S., et al. (2001). Preseismic deformation and coseismic displacements associated with the 1999 Chi–Chi, Taiwan, earthquake. *Bulletin of the Seismological Society of America, 91*(5), 995–1012.

(Received March 24, 2017, revised August 4, 2017, accepted September 8, 2017)

Pure Appl. Geophys.
© 2017 Springer International Publishing AG, part of Springer Nature
https://doi.org/10.1007/s00024-017-1737-1

# Anomalous Streamflow and Groundwater-Level Changes Before the 1999 M7.6 Chi–Chi Earthquake in Taiwan: Possible Mechanisms

CHI-YU KING[1] and YEEPING CHIA[2]

*Abstract*—Streamflow recorded by a stream gauge located 4 km from the epicenter of the 1999 M7.6 Chi–Chi earthquake in central Taiwan showed a large and rapid anomalous increase of 124 $m^3$/s starting 4 days before the earthquake. This increase was followed by a comparable co-seismic drop to below the background level for 8 months. In addition, groundwater-levels recorded at a well 1.5 km east of the seismogenic fault showed an anomalous rise 2 days before the earthquake, and then a unique 4-cm drop beginning 3 h before the earthquake. The anomalous streamflow increase is attributed to gravity-driven groundwater discharge into the creek through the openings of existing fractures in the steep creek banks crossed by the upstream Shueilikun fault zone, as a result of pre-earthquake crustal buckling. The continued tectonic movement and buckling, together with the downward flow of water in the crust, may have triggered the occurrence of some shallow slow-slip events in the Shueilikun and other nearby fault zones. When these events propagate down-dip to decollement, where the faults merges with the seismogenic Chelungpu fault, they may have triggered other slow-slip events propagating toward the asperity at the hypocenter and the Chelungpu fault. These events may then have caused the observed groundwater-level anomaly and helped to trigger the earthquake.

## 1. Introduction

Many earthquake-triggered groundwater-level and streamflow changes have been observed since the 1960s (Waller 1966; Roeloffs 1998; King et al. 1999; King and Igarashi 2002; Manga et al. 2003; Montgomery and Manga 2003). Whereas the groundwater-level increases and decreases occurred mostly coseismically, the streamflow changes usually

**Electronic supplementary material** The online version of this article (https://doi.org/10.1007/s00024-017-1737-1) contains supplementary material, which is available to authorized users.

[1] Earthquake Prediction Research, Inc, Los Altos, CA, USA.
E-mail: chiyuking@aol.com
[2] Department of Geosciences, National Taiwan University, Taipei, Taiwan, ROC.

happened shortly after the earthquakes, and they were mostly increases (Manga et al. 2003; Muir-Wood and King 1993; Wang et al. 2003, 2004; Wang and Manga 2010). These changes have generally been attributed to earthquake-related changes in crustal strain and in hydrogeological properties, such as permeability. In this study, we try to better understand the mechanisms of a streamflow change and a groundwater-level change recorded near the epicenter before and after the 1999 M7.6 Chi–Chi earthquake in central Taiwan.

Taiwan is located in a subduction zone, where the Philippine Sea plate subducts northward beneath the Eurasian plate along the Ryukyu Trench in the north and the Eurasian plate subducts eastward beneath the Philippine Sea plate along the Manila Trench in the south. The plate convergence rate is about 8 cm/year with a contraction rate of 2.0–4.5 cm/year in the mountainous fold-and-thrust belt of western Taiwan, where the Chi–Chi earthquake occurred (Yu et al. 2001).

The Chi–Chi earthquake occurred at 1:47 am on 21 September 1999 local time (17:47 on 20 September UTC), at the end of a rainy season, with an epicenter located at 23.85°N; 120.82°E and a depth of approximately 8 km (Fig. 1). The focal mechanism was oblique thrust with a strike of 5°, a dip of 34° and a rake of 65°. Co-seismic surface ruptures, extending approximately 100 km, coincide with the trace of the north–south trending Chelungpu thrust fault in the western Taiwan fold-and-thrust foothills area. GPS data showed a co-seismic horizontal offset of 2.4–10.1 m across the Chelungpu fault, and a vertical offset of 1.2–4.4 m. The crustal deformation before the earthquake was essentially a uniaxial compression in the direction of 114° at a

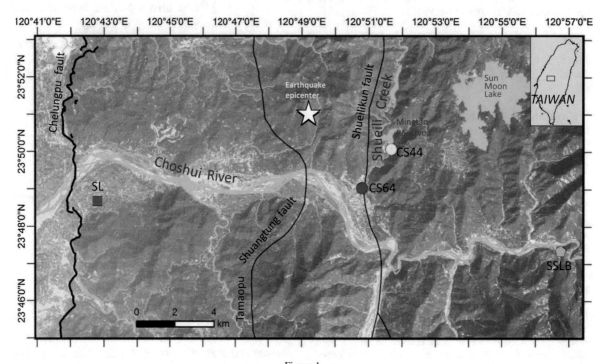

Figure 1
Location of monitoring well SL (blue square), stream gauges CS64 (red circle) and CS44 (yellow circle), and seismic station SSLB (green pentagons), along with the Shueili Creek, Choshui River, Mingtan reservoir, Sun Moon Lake and earthquake epicenter (white star) in central Taiwan. Pre-earthquake streamflow and groundwater-level anomalies were observed at CS64 and SL, respectively (Modified from Liu et al. this issue)

strain rate of 0.36 μ/year (Yu et al. 2001). After the earthquake, most nearby buildings on the hanging wall of the fault were seriously damaged, but very few on the footwall were, even located only a few meters away from the surface rupture (Chen et al. 2000; Dong et al. 2003).

## 2. Observations

### 2.1. Anomalous Streamflow Changes

Streamflow discharges were measured at 32 gauges on four major rivers (Daan River, Dajia River, Wu River, and Choshui River) and their tributaries (refer to ESM 1). Most of them are located on the mountainous east side of the Chelungpu fault. These gauges were installed at the riverside or on the bridge that crosses the river where the water level was measured. The water level is recorded and converted to the discharge automatically by a data logger. Among the 23 of them that recorded clear

earthquake-related anomalies, 22 were post-earthquake streamflow increases, reaching their peak values 1–4 days after the earthquake (Liu et al. this issue).

The change recorded by the remaining gauge, CS64, is quite different. This gauge is the closest to the epicenter, approximately 4 km to the southeast, and almost directly above the hypocenter. It is located downstream of the southwestward-flowing Shueili Creek, a major tributary of the Choshui River (Fig. 1). Immediate upstream, the creek is in a deep valley and is crossed obliquely by the north–south trending Shueilikun fault zone, which is parallel to and on the east side of the seismogenic Chelungpu and the Tamaopu Shuangtung faults. These faults are part of the western Taiwan fold-and-thrust system, and they merge at a depth of about 5.5 km or greater (Fig. 6b). In the vicinity of CS64, the Shueili Creek watershed is located on the Shiuchangliu formation which is composed of fractured quartzitic sandstone and arkosic sandstone inter-bedded with argillite (Fig. 2).

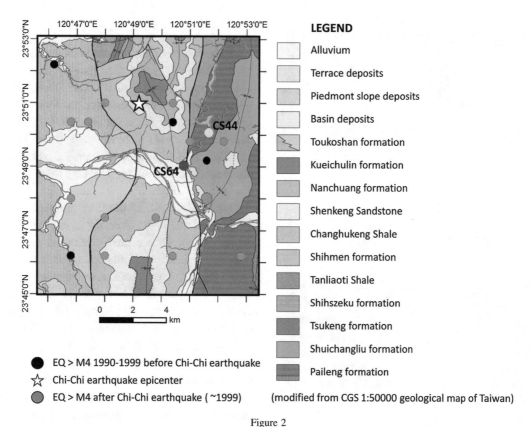

Figure 2
Geologic map with active fault surface traces near the epicenter of Chi–Chi earthquake (white star). Earthquakes between 1990 and 1999 with magnitude greater than 4 are presented, black circles represent the earthquakes occurred before Chi–Chi earthquake and gray circles represent the aftershocks

Four days before the earthquake, the discharge at CS64 showed a large and rapid increase from 47 to 171 $m^3/s$ in 2 days, and then a comparably large and rapid post-earthquake drop in 2 days to about 7 $m^3/s$, which is much lower than the pre-earthquake average level of about 50 $m^3/s$ (Fig. 3a). The post-earthquake streamflow remained at a level below 33 $m^3/s$ till the end of 1999. Without heavy rainfall and similar increases at other nearby gauges, including CS44, which was only about 3 km upstream (Fig. 1), this pre-earthquake increase is obviously not due to rainfall.

Figure 3b shows the data recorded at CS44, which was established at the base of the Mingtan dam for the purpose of monitoring reservoir release to the Shueili Creek. Usually the discharge observed here showed lesser amount than, but in similar patterns as, that at CS64. But, for this earthquake CS44 did not show any pre-seismic increase or post-seismic decrease as recorded at CS64. Instead, it showed a jump to 90 $m^3/s$ from the pre-earthquake level of 14 $m^3/s$ 2 days after the earthquake. However, this increase was found later to be caused by water release from the reservoir for generating additional electric power to meet the emergency need and dam safety in the earthquake's aftermath. Even with such an increase, the larger streamflow did not cause any increase at CS64, which is only 3-km downstream. Instead, the post-earthquake discharge at CS64 actually decreased, as described above.

Figure 3c compares the discharge time series of CS64 and CS44 over a longer period of 3 years. Prior to the earthquake, they showed similar patterns of variation, except that the discharge at CS64 downstream was larger, as expected. After the earthquake, however, the situation reversed for about 8 months.

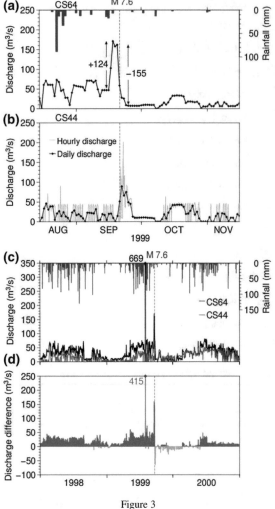

## Figure 3

Stream discharges recorded at gauges on the Shueili Creek near the earthquake epicenter **a** Daily discharges at CS64 from August to November 1999 (raw data are listed in ESM 2); **b** Daily and hourly discharges at CS44 from August to November 1999; **c** Comparison of daily discharges at CS64 (black line) and CS44 (red line) from 1998 to 2000; **d** Difference in daily discharges between CS64 and CS44. Earthquake time is indicated by the vertical dash line (Modified from Liu et al. this issue)

In other words, the nature of Shueili Creek changed coseismically from water-gaining to water-losing (Fig. 3d). This unprecedented anomalous phenomenon indicates that, during this 8 month, the creek water in the 3-km section between the two stations crossed by the Shueilikun fault was continuously leaking through the creek bed down to the crust (Fig. 2).

## 2.2. Anomalous Groundwater-Level Changes

Sustained co-seismic water-level changes induced by the Chi–Chi earthquake, ranging from a fall of 11.09 m to a rise of 7.42 m, were observed at 152 wells located on the west side, or footwall, of the Chelungpu fault (Chia et al. 2001, 2008). The majority of these changes were rises, but those recorded at 14 wells in a narrow zone adjacent to the ruptured Chelungpu fault were falls. Sustained changes are attributable to static-strain changes in the crust as a result of the fault movement (Wakita 1975; Roeloffs 1988; Grecksch et al. 1999; Ge and Stover 2000). Others have attributed decreases to shaking induced crustal dilatation (Mohr et al. 2015).

A unique pre-earthquake groundwater-level change was recorded, however, at SL which was the only monitoring well situated in the hanging wall of the seismogenic fault (Fig. 4). SL well is located 1.5 km east of the surface rupture of the fault and approximately 1 km above the fault plane. The well tapped a shallow unconfined aquifer, and thus the water level does not change with either barometric pressure or earth tide (Hussein et al. 2013; Jacob 1940). The well is 24 m deep with a screen at depth between 9 and 18 m. The aquifer is primarily composed of unconsolidated gravel and coarse sand, which is underlain by bedrock below the depth of 18 m and overlain by a thin layer of less permeable silty clay. Water level was measured by a pressure transducer and recorded by a data logger at the wellhead. The groundwater level, ranging between 4 and 9 m below the ground surface (Fig. 5a), generally declines slowly, at a rate of 0–1 cm/hr. Over a period of 3 h, it usually drops 0–2 cm (Fig. 5b). It varies significantly only with heavy rainfall and local irrigation from surface water. Between January 1997 and November 2000, however, there were no pumping or injection activities in the vicinity of the well.

As shown in Fig. 5c, the groundwater-level, which had been declining since the beginning of September at the end of the rainy season, turned around anomalously on September 19 (2 days before the earthquake) to show a rising trend, without any significant rainfall. The rise lasted until 3 h before the earthquake, when the water level began to show a rapid pre-earthquake drop of 4 cm (Fig. 5d). There

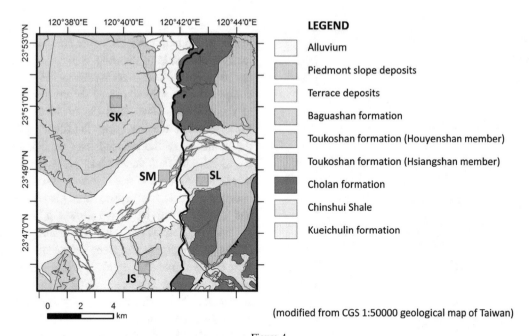

Figure 4
Geologic map with active fault surface traces and the surface rupture of Chelungpu Fault. The four groundwater monitoring wells (squares) are located on both side of the rupture. SK, SM and JS are on the footwall and SL is on the hanging wall of the Chelungpu fault

was a 14-cm coseismic spike, which was probably induced by the passing seismic waves, and is thus not considered further. The decline continued at a slower rate after the earthquake. Since the water level at SL had never declined more than 3 cm during the initial 3 h of a falling trend (Chia et al. 2008), this 4-cm pre-earthquake decline (Fig. 5b), together with the preceding 2-day increase, is considered not random but a unique earthquake-related anomaly.

Groundwater-levels were also recorded in 9 well at 5 stations near the SL well (Fig. 4). These wells are all located on the west side, or footwall, of the seismogenic fault. Unlike the SL well, they did not show any anomalous groundwater-level changes before the earthquake.

## 3. Conceptual Model

Gauge CS64 and well SL are both situated near the epicenter on the hanging wall of the seismogenic Chelungpu fault. To explain the large and rapid streamflow anomaly at CS64 and the unique groundwater-level anomaly at SL, we propose a

macroscopic model that takes the heterogeneity of the earth's crust into account. On the hanging wall of this thrust fault, many large fractures and macroscopic fissures had been generated by past tectonic movements and earthquakes (Lee and Chan 2007). In fracture mechanics, it is well known that, when a brittle solid is increasingly stressed to the critical close-to-failure point, stress tends to concentrate on some pre-existing weak parts (crack tips), before the major failure begins from the weakest (Hertzberg 1996). The same should be true also for the earth's crust in the case of a large earthquake. Before the earthquake, the crustal stress tends to concentrate in the seismogenic and nearby fault zones to induce small movements around the asperities, the strongest of which is located at the hypocenter, causing crustal and surface fluids to migrate. The anomalous hydrologic changes recorded at CS64 and SL near the epicenter may be a manifestation of such a process.

### 3.1. Streamflow

The large and rapid streamflow increase recorded at CS64 4 days before the earthquake was not a

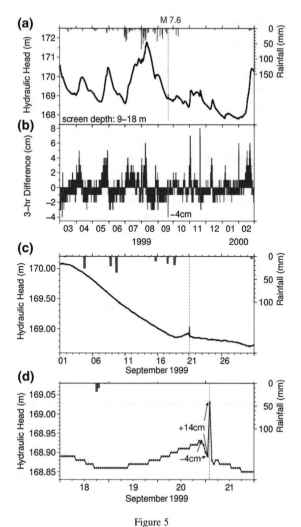

Figure 5
Ground water levels recorded at the near-epicenter well SL.
a Hourly data at SL from March 1999 to February 2000 showing
the variation of groundwater-level (raw data are listed in ESM 3)
and rainfall. b Groundwater-level change over a 3-h period. The
4-cm drop during the initial 3-h period in the recession stage
immediately before the earthquake is a unique phenomenon.
c Groundwater-level at SL in September 1999 showing an
anomalous increase 2 days before the earthquake. d Anomalous
groundwater-level drop of 4 cm that began 3 h before the
earthquake

regular creek-flow increase, because no such increase
was observed at CS44 about 3 km upstream. It must
have been caused by groundwater discharge from
higher grounds of the creek banks in the 3 km section
between these two gauges, especially in the north–
south trending Shueilikun fault zone that crosses this
section obliquely. Prior to the Chi–Chi earthquake,
the fault movement must have been stuck at the

seismogenic asperity near the hypocenter in the fault
plane, which is directly under the study area, and the
hanging wall side of the crust was under an increasing tectonic sideway compression (Fig. 6b). This
compression may have caused the crust to buckle up
in the epicentral area, and the buckling may have
caused surface tension under the higher grounds of
the mountain in the buckled area, especially within
the Shueilikun fault zone. Many pre-existing vertical
cracks and fissures in the area may have opened up
because of their low tensile strength. These cracks
and fissures in turn may have allowed a large amount
of stored groundwater after the rainy season to
discharge rapidly to the steep banks of the Shueili
Creek down into the creek, causing the observed
rapid streamflow increase at CS64 4 days before the
earthquake. No such flow increase was observed in
the neighboring Tamaopu Shuangtung fault zone that
crosses the larger Choshui river, or elsewhere,
probably because of lack of observation or/and their
being situated on flatter land. The subsequent large
and rapid co-seismic and post-seismic streamflow
decrease was probably caused by crustal extension
and flattening of the hanging wall to a depth of
decollement (Fig. 6b) during and after the fault
rupture. The crustal extension then opened up many
vertical cracks and fissures in the hanging wall,
including those in the Shueili streambed. Through
such openings huge amount of creek water could
discharge rapidly at a rate of approximately $3 \times 10^{-4}$
m/s to the bedrock beneath the creek, assuming the
width of creek valley floor to be 100 m. The period of
8 months for the streamflow to return to normal is
probably how long these cracks and fissures needed
to heal (Li et al. 2006). Post-earthquake field
inspections indeed found vertical tensile cracks in
many areas on the hanging wall of the fault (Angelier
et al. 2000; Lee et al. 2000, 2002; Wang et al. 2016).

### 3.2. Groundwater-Level

The above-mentioned crustal disturbance, including buckling and water flow may have caused some
slow-slip events to occur along the Shueilikun fault
and the neighboring Tamaopu Shuangtung fault,
which is crossed by Choshui River (Fig. 1). These
events may have then propagated down-dip at a speed

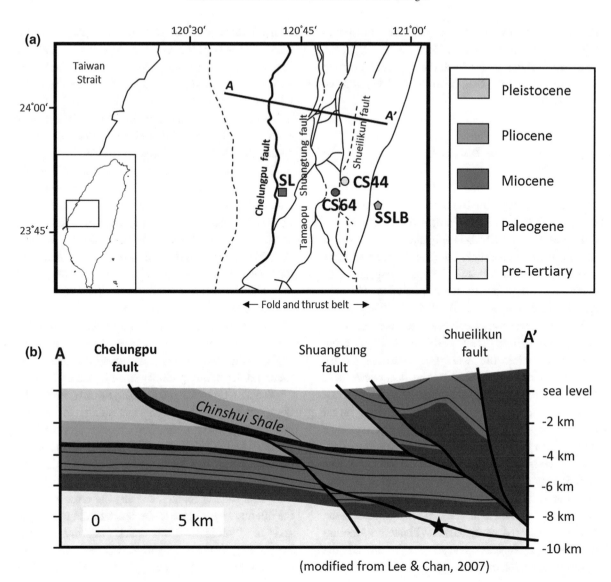

Figure 6
Geological structure of the central western Taiwan. **a** Planar view of geological structure near the Chelungpu fault and the locations of SL, CS64, CS44 and SSLB. **b** Interpreted geological cross-section of A–A' showing the structures of the Chelungpu fault area (Lee and Chan 2007). The hypocenter of Chi–Chi earthquake is indicated by a star

of up to 10 km/day (King 1973) to the decollement, where the faults merges with the Chelungpu fault, and from there triggered other slow-slip events to propagate toward the asperity at the hypocenter and the Chelungpu fault (Fig. 6b). An up-dip propagating event toward the surface trace of the Chelungpu fault may have caused a load transfer in the near-surface layer, causing a pore-volume contraction and the observed anomalous water-level rise at SL 2 days

before the earthquake, as it was approaching the well. Similarly, it may have caused a pore-volume extension and the 4-cm water-level drop 3 h before the earthquake, when it was departing from the well. The static volumetric strain corresponding to the anomalous 4-cm water-level change is estimated to be about $3 \times 10^{-9}$–$9 \times 10^{-8}$ (refer to ESM 4). This, together with the reasonably assumed rupture length of 8–16 km and duration of 2–3 days, gives an estimated moment magnitude value of 4–5 for the corresponding slow-slip event (King 1973; Linde et al. 1996). The load shift to the asperity at the hypocenter caused by these events may then have finally triggered the earthquake.

## 4. Discussion

This conceptual model is admittedly speculative, and there is no geodetic data to test it. However, there was some broadband-seismic data record at two stations, one of which SSLB is only about 10 km east of CS64 (Fig. 1). In the data, Lin (2012) observed some small anomalous crustal vertical displacements beginning 5 days before the earthquake. The first displacement was an uplift of about 0.48 mm at SSLB, and it was followed by several small elevation drops, ranging from 0.22 to 0.40 mm at SSLB during the 4 days before the earthquake. Assuming these displacements to be caused by the occurrence of a series of slow-slip events along the decollement about 10–12 km deep (Fig. 6b), he attributed the initial uplift to a slow-slip event with a slip of 16 cm along a rupture area of 10 km × 10 km about 25 km to the east of SSLB. This uplift is consistent with our buckling model, and this slow-slip event may have been the "last straw" to cause the proposed opening up of cracks/fissures in the Shueilikun fault zone 4 days before the earthquake. Without considering possible hydrological changes, he attributed the observed elevation drops during a period of 3–4 days before the earthquake to slow-slip events of revered direction along the decollement to the east of SSLB. Since reversed slip is tectonically unreasonable, we think the elevation drop may be more likely caused by the water loss from the crust to the streams, as described above. He also attributed the elevation drops during the last 2 days before the earthquake to two other slow-slip events along the decollement near the Chelungpu fault. This suggestion is consistent with our model in explaining the anomalous water-level changes at the SL well.

Fault-gouge layers needed for slow-slip to occur along the Chelungpu fault plane were observed in cores obtained from drilled holes (Hung et al. 2009). The possibility of slow-slip occurrence along the fault is also indicated by the observations that post-earthquake slip was partly aseismic (Hsu et al. 2002) and that during the earthquake the amount of seismic slip was relatively small compared with the aseismic slip (Wu et al. 2001; Ma et al. 2003), especially in the epicentral area (Johnson et al. 2001). Water-level changes associated with aseismic creep have been studied previously (Roeloffs et al. 1989).

The occurrence time (few days to several weeks before the earthquake) of the presently observed anomalies is consistent with that of some other observed anomalies, such as geomagnetic field (Yen et al. 2004) and ionospheric total electron content (Lin 2013), although the validity of such observations have been questioned (Doglioni et al. 2013; Masci and Thomas 2015; Masci et al. 2015, 2017). The proposed pre-earthquake crack/fissure opening in the Shueilikun fault zone, and thus the detachment of the hanging wall of the Chelungpu thrust fault, may explain why the coseismic slip along the Chelungpu fault and the building damages on it was so large.

The above-proposed triggering mechanism for the observed anomalies is somewhat similar to the "dilatancy and fluid migration" model proposed by Scholz et al. (1973) and Nur (1974): Prior to an earthquake, as the local stress is approaching the critical level, small cracks and fissures occur in the crust near the hypocenter, causing the crust to dilate; the dilation in turn causes the crustal fluid to migrate and to "lubricate" the stuck asperity at the hypocenter, and subsequently leads to its failure. In the present case, however, we have also invoked crustal buckling, slow-slip events as well as downward water flow by gravity. Enhancement of vertical permeability of the groundwater system during the Chi–Chi earthquake was proposed previously by Wang et al. (2004) also.

The existence of pre-earthquake anomalies, if confirmed by more field data, should encourage earthquake scientists to resume efforts of pursuing short-term earthquake prediction, which currently is deemed impossible by many (e.g., Geller et al. 1997). To be successful in such attempts, it is important to recognize the heterogeneity of the crust and to install appropriate monitoring instruments at a sufficient number of carefully selected "sensitive" sites located along some weak/fault zones, where the crustal stress is concentrated.

## Acknowledgement

The authors would like to thank two reviewers for their thoughtful comments. We gratefully acknowledge access to hydrological data of the Water Resources Agency of Taiwan and the Taiwan Power Company. This work is supported by the Ministry of Science and Technology of Taiwan (MOST 104-2116-M-002-008) and the Industrial Technology Research Institute. Special thanks are extended to many colleagues for collecting and processing data used in this paper.

## REFERENCES

Angelier, J., Chu, H. T., Lee, J. C., Hu, J.C., Mouthereau, F., Lu, C. Y., et al. (2000). Geologic knowledge and seismic risk mitigation: Insight from the Chi-Chi earthquake (1999), Taiwan. In C.-H. Lo & W.-I. Liao (Eds.), *Proceedings of international workshop on annual commemoration of Chi-Chi earthquake, science aspect* (pp. 13–24).

Chen, C. C., Huang, C. T., Cherng, R. H., & Jeng, V. (2000). Preliminary investigation of damage to near fault buildings of the 1999 Chi-Chi earthquake. *Earthquake Engineering and Engineering Seismology, 2*(1), 79–92.

Chia, Y., Chiu, J. J., Chiang, Y. H., Lee, T. P., & Liu, C. W. (2008). Spatial and temporal changes of groundwater level induced by thrust faulting. *Pure and Applied Geophysics, 165*(1), 5–16. https://doi.org/10.1007/s00024-007-0293-5.

Chia, Y. P., Wang, Y. P., Chiu, J. J., & Liu, C. W. (2001). Changes of groundwater level due to the 1999 Chi-Chi earthquake in the Choshui River alluvial fan in Taiwan. *Bulletin of the Seismological Society of America, 91*(5), 1062–1068. https://doi.org/10.1785/0120000726.

Doglioni, C., Barba, S., & Carminati, E. (2013). Fault on-off versus strain rate and earthquakes energy. *Geoscience Frontiers, 6*(2), 265–276. https://doi.org/10.1016/j.gsf.2013.12.007.

Dong, J. J., Wang, C. D., Lee, C. T., Liao, J. J., & Pan, Y. W. (2003). The influence of surface ruptures on building damage in the 1999 Chi-Chi earthquake: A case study in Fengyuan City. *Engineering Geology, 71*, 157–179.

Ge, S., & Stover, S. C. (2000). Hydrodynamic response to strike- and dip-slip faulting in a half space. *Journal of Geophysical Research, 105*(B11), 25513–25524. https://doi.org/10.1029/2000JB900233.

Geller, R. J., Jackson, D. D., Kagan, Y. Y., & Mulargia, F. (1997). Can earthquakes be predicted? *Science, 278*, 488–490.

Grecksch, G., Roth, F., & Kümpel, H. J. (1999). Co-seismic well-level changes due to the 1992 Roermond earthquake compared to static deformation of half-space solutions. *Geophysical Journal International, 138*(2), 470–478. https://doi.org/10.1046/j.1365-246X.1999.00894.x.

Hertzberg, R. W. (1996). *Deformation and fracture mechanics of engineering materials.* Wiley.

Hsu, Y.-J., Boehor, N., Segall, P., Yu, S.-B., & Kuo, L.-C. (2002). Rapid afterslip following the 1999 Chi-Chi, Taiwan, earthquake. *Geophysical Research Letters, 29*(16), 1–4. https://doi.org/10.1029/2002GL014967.

Hung, J., Ma, K., Wang, C., Ito, H., Lin, W., & Ye, E. (2009). Subsurface structure, physical properties, fault-zone characteristics and stress state in scientific drill holes of Taiwan Chelungpu fault drilling project. *Tectonophysics, 466*, 307–321.

Hussein, M., Odling, N., & Clark, R. (2013). Borehole water level response to barometric pressure as an indicator of aquifer vulnerability. *Water Resources Research, 49*(10), 7102–7119. https://doi.org/10.1002/2013WR014134.

Jacob, C. E. (1940). The flow of water in an elastic artesian aquifer. *Eos Transactions AGU, 21*, 574. https://doi.org/10.1029/TR021i002p00574.

Johnson, K. M., Hsu, Y. J., Segall, P., & Yu, S. B. (2001). Fault geometry and slip distribution of the1999 Chi-Chi, Taiwan earthquake imaged from inversion of GPS data. *Geophysical Research Letters, 28*, 2285–2288.

King, C.-Y. (1973). Kinematics of fault creep. *Philosophical Transactions of the Royal Society of London A, 274*, 355–360.

King, C.-Y., Azuma, S., Igarashi, G., Ohno, M., Saito, H., & Wakita, H. (1999). Earthquake-related water-level changes at 16 closely clustered wells in Tono, central Japan. *Journal of Geophysical Research, 104*(B6), 13073–13082.

King, C.-Y., & Igarashi, G. (2002). Earthquake related hydrologic and geochemical changes. *International Handbook of Earthquake and Engineering Seismology, 81A*, 637–645.

Lee, J. C., & Chan, Y. C. (2007). Structure of the 1999 Chi-Chi earthquake rupture and interaction of thrust faults in the active fold belt of western Taiwan. *Journal of Asian Earth Sciences, 31*(3), 226–239. https://doi.org/10.1016/j.jseaes.2006.07.024.

Lee, J. C., Chu, H.-T., Angelier, J., Chan, Y.-C., Hu, J.-C., Lu, C.-Y., et al. (2002). Geometry and structure of northern surface ruptures of the 1999 $M_w = 7.6$ Chi-Chi Taiwan earthquake: Influence from inherited fold belt structure. *Journal of Structural Geology, 24*, 173–192.

Lee, C. T., Kelson, K. I. & Kang, K. H. (2000). Hanging wall deformation and its effect on buildings and structures as learned from the Chelungpu fault in the 1999 Chi-Chi Taiwan earthquake. In C.-H. Lo & W.-I. Liao (Eds.), *Proceeding of international workshop on annual commemoration of Chi-Chi earthquake* (pp. 93–104). Science Aspect.

Li, Y. G., Chen, P., Cochran, E., Vidale, J., & Burdette, T. (2006). Seismic evidence for rock damage and healing on the San Andreas fault associated with the 2004 M 6.0 Parkfield Earthquake. *Bulletin of the Seismological Society of America, 96*(4B), S349–S363. https://doi.org/10.1785/0120050803.

Lin, C.-H. (2012). The possible observation of slow slip events prior to the occurrence of the 1999 Chi-Chi earthquake. *Terrestrial, Atmospheric and Oceanic Sciences, 23*, 145–159.

Lin, J.-W. (2013). Taiwan's Chi-Chi earthquake precursor detection using nonlinear principal component analysis to multichannel total electron content records. *Journal of Earth Science, 24*(2), 244–253.

Linde, A. T., Gladwin, M., Johnston, M., Gwyther, R., & Bilham, R. (1996). A slow earthquake sequence on the San Andreas Fault. *Nature, 383,* 65–68.

Liu, C.-Y., Chia, Y., Chuang, P.-Y., Ge, S., Wang, C. Y. & Teng, M.-H. (2017). Streamflow changes in the vicinity of seismogenic fault induced by the 1999 Chi-Chi earthquake. *Pure and Applied Geophysics* (this issue).

Ma, K. F., Brodsky, E. E., Mori, J., Ji, C., & Song, T. R. A. (2003). Evidence for fault lubrication during the 1999 Chi-Chi, Taiwan, earthquake (Mw 7.6). *Geophysical Research Letters, 30*(5), 1244. https://doi.org/10.1029/2002GL015380.

Manga, M., Brodsky, E. E., & Boone, M. (2003). Response of streamflow to multiple earthquakes. *Geophysical Research Letters, 30*(5), 1214. https://doi.org/10.1029/2002GL016618.

Masci, F., & Thomas, J. N. (2015). Are there new findings in the search for ULF magnetic precursors to earthquakes? *Journal of Geophysical Research: Space Physics, 120,* 10289–10304. https://doi.org/10.1002/2015JA021336.

Masci, F., Thomas, J. N., & Secan, J. A. (2017). On a reported effect in ionospheric TEC around the time of the 6 April 2009 L'Aquila earthquake. *Natural Hazards Earth Sustainable Science, 17,* 1361–1468. https://doi.org/10.5194/nhess-17-1461-2017.

Masci, F., Thomas, J. N., Villani, F., Secan, J. A., & Rivera, N. (2015). On the onset of ionospheric precursors 40 min before strong earthquakes. *Journal of Geophysical Research: Space Physics, 120,* 1383–1393. https://doi.org/10.1002/2014JA020822.

Mohr, C., Manga, M., Wang, C. Y., Kirchner, J., & Bronstert, A. (2015). Shaking water out of soil. *Geology, 43*(3), 207–210. https://doi.org/10.1130/G36261.1.

Montgomery, D. R., & Manga, M. (2003). Streamflow and water well responses to earthquakes. *Science, 300*(5628), 2047–2049. https://doi.org/10.1126/science.1082980.

Muir-Wood, R., & King, G. C. P. (1993). Hydrologic signatures of earthquake strain. *Journal of Geophysical Research, 98*(B12), 22035–22068.

Nur, A. (1974). Matsushiro, Japan, earthquake swarm: Confirmation of the dilatancy-fluid diffusion model. *Geology, 2*(5), 217–221.

Roeloffs, E. A. (1988). Hydrologic precursors to earthquakes: a review. *Pure and Applied Geophysics, 126*(2–4), 177–209. https://doi.org/10.1007/BF00878996.

Roeloffs, E. A. (1998). Persistent water level changes in a well near Parkfield, California, due to local and distant earthquakes. *Journal of Geophysical Research, 103*(B1), 869–889.

Roeloffs, E. A., Burford, S. S., & Riley, F. S. (1989). Hydrologic effects on water level changes associated with episodic fault creep near Parkfield, California. *Journal of Geophysical Research, 94,* 12387–12402.

Scholz, C. H., Sykes, L. R., & Aggarwal, Y. P. (1973). Earthquake prediction: A physical basis. *Science, 181,* 803–810.

Wakita, H. (1975). Water wells as possible indicators of tectonic strain. *Science, 189*(4202), 553–555. https://doi.org/10.1126/science.189.4202.553.

Waller, R. M. (1966). *Effects of the March 1964 Alaska earthquake on the hydrology of Anchorage area.* Prof. Pap. 544B, USGS.

Wang, C. Y., Dreger, D. S., Wang, C. H., Mayeri, D., & Berryman, J. G. (2003). Field relations among coseismic ground motion, water level change and liquefaction for the 1999 Chi-Chi (MW = 7.5) earthquake, Taiwan. *Geophysical Research Letters, 30*(17), 1890. https://doi.org/10.1029/2003GL017601.

Wang, C. Y., Liao, X., Wang, L.-P., Wang, C.-H., & Manga, M. (2016). Large earthquakes create vertical permeability by breaching aquitards. *Water Resources Research.* https://doi.org/10.1002/2016WR018893.

Wang, C. Y. & Manga, M. (2010). Increased stream discharge (Ch. 4). In *Earthquakes and water* (pp. 45–66). Springer.

Wang, C. Y., Manga, M., Dreger, D., & Wong, A. (2004). Streamflow increase due to rupturing of hydrothermal reservoirs: evidence from the 2003 San Simeon, California, earthquake. *Geophysical Research Letters, 31*(10), L10502. https://doi.org/10.1029/2004GL020124.

Wu, C., Takeo, M., & Ide, S. (2001). Source process of the Chi-Chi earthquake: A joint inversion of strong motion data and global positioning system data with a multifault model. *Bulletin of the Seismological Society of America, 91*(5), 1128–1143.

Yen, H.-Y., Chen, C.-H., Yeh, Y.-H., Liu, J.-Y., & Lin, C.-R. (2004). Geomagnetic fluctuation during 1999 Chi-Chi earthquake in Taiwan. *Earth, Planets and Space, 56,* 39–45.

Yu, S.-B., Kuo, L. C., Hsu, Y. J., Su, H. H., Liu, C. C., Hou, C. S., et al. (2001). Preseismic deformation and coseismic displacements associated with the 1999 Chi-Chi, Taiwan, earthquake. *Bulletin of the Seismological Society of America, 91,* 5–1012.

(Received April 1, 2017, revised November 23, 2017, accepted November 27, 2017)

Pure Appl. Geophys.
© 2017 Springer International Publishing
DOI 10.1007/s00024-017-1579-x

# Preseismic Changes of Water Temperature in the Yushu Well, Western China

XIAOLONG SUN,[1] YANG XIANG,[2] ZHEMING SHI,[3] (iD) and BO WANG[4]

*Abstract*—We observed abnormal changes of the water temperature in Yushu (YSWT) well, China, most of which were followed by earthquakes. This study statistically analyzes the correlation between the magnitude and duration of the anomalies in YSWT and earthquakes in the Tibetan block and its margins. The effectiveness of using observed YSWT data to predict earthquakes was quantitatively examined by the Molchan error diagram method. The results show that (1) the YSWT underwent several abnormal changes marked by "V"-shaped patterns, which might be related to several earthquakes that occurred in the Tibetan block and its margins. The extent and duration of the abnormal changes in the YSWT were linearly related to the magnitude of the earthquake; i.e., the higher the magnitude, the greater the change in the YSWT, and the shorter the duration. (2) Abnormal changes in the YSWT are somewhat predictive of earthquakes with magnitudes $\geq 5.5$ ($\geq M5.5$) within 800 km of the Yushu well and $\geq M6.5$ earthquakes in the Tibetan block and its margins. The prediction has a probability gain of approximately 2, and the most likely time period for an earthquake to occur is within approximately 3 months after the occurrence of an YSWT anomaly. Most of the anomalies in YSWT appeared before earthquakes in the thrust of block margins. Notably the larger strains from the earthquake did not produce any response. We speculate that the preseismic responses reflect the regional tectonics, such as the motion of the Indian plate, straining sub-blocks of the Tibetan block.

**Key words:** Preseismic changes, Water temperature, Yushu well, Molchan error diagram.

## 1. Introduction

Groundwater in the crust can be very sensitive to reflect changes in crustal stress and hence may play a vital role in geodynamical processes such as subsurface heat transfer, tectonic and seismic activity (Che et al. 1998; Du 2007; Faulkner and Rutter 2003; Caine and Minor 2009). Changes in well water temperatures are a response of groundwater observation systems to crustal deformation and the migration of groundwater (Asteriadi and Liverato 1989; Kitagawa et al. 1996). Observations of changes in well water temperatures are one of the most direct ways of revealing the pattern of the responses to stresses (Shi and Wang 2014). Investigations of earthquake-induced water temperature changes in the near field have practical significance, including forecasting earthquake-affected water supplies (Roeloffs 1998), understanding crustal deformation (Johnston et al. 1995), and studying the maintenance of permeability in enhanced geothermal systems (Manga et al. 2012) and may shed light on the mechanisms of possible hydrogeological responses to earthquakes (Wakita 1975; Simpson 1986; Montgomery and Manga 2003; Wang et al. 2012, 2013).

Water temperature changes in wells following earthquakes have been widely documented (Mogi et al. 1989; Shimamura et al. 1984; Kitagawa et al. 1996; Shi and Wang 2014; Wang et al. 2012). These studies show that water temperature changes can be caused by changes in water level or crustal strain induced by earthquakes. Therefore, aquifer dilatation due to earthquakes may cause changes in water level and water temperature (Sun and Liu 2012). Sadovsky et al. (1972) reported increases in the water level and temperature in a well located 30 km from the epicenter prior to the main shock of the 1970 M6.7 Przhevalsk earthquake in Kyrgyzstan. Wakita (1975) observed changes in water wells as possible indicators of tectonic strain associated with earthquakes. Shi et al. (2015) suggested that permeability change is a more plausible mechanism than static strain to explain most of the co-seismic changes associated

---
[1] Key Laboratory of Crustal Dynamics, Institute of Crustal Dynamics, China Earthquake Administration, Beijing 100085, China.

[2] Earthquake Administration of Xinjiang Uygur Autonomous Region, Urumqi 830002, China.

[3] School of Water Resources and Environment, China University of Geosciences, Beijing 100083, China. E-mail: szm@cugb.edu.cn

[4] China Earthquake Networks Center, Beijing 100045, China.

with earthquakes. As water temperature changes imply of groundwater migration, water temperature records accompanying water-level data are useful for examining the mechanisms of hydrological changes.

Located in the southwestern mainland China, the Tibetan Plateau is the largest active seismic zone in mainland China and is marked by the most intense seismic activity and frequent large earthquakes. Significant water temperature anomalies in the Yushu well with similar patterns were recorded before several large earthquakes in the Tibetan block and its margins. The study of He et al. (2012) focused on the anomaly characteristics before Wenchuan M7.9 earthquake, and Yang et al. (2016) focused on the anomaly characteristics before Nepal M7.8 earthquake. The present study will focus on (1) the similarities and differences in the anomalies of the water temperature in the Yushu well by investigating the relationship between the characteristics of pre-earthquake changes in the YSWT and large earthquakes in the nearby regions; (2) quantitative analysis of the effectiveness of the YSWT data to predict earthquakes using the Molchan error diagram method.

## 2. Observational Background

The Yushu observation well (coordinates: 97.02°E, 33.01°N) is located in Tuanjie Village, Jiegu Township, Yushu County in the Chinese province of Qinghai at the southern foot of the Bayan Har Mountains and the eastern end of the Tanggula Mountains. Most of the faults in this region are located along west-northwest-trending arc-shaped strike-slip faults that protrude toward the northeast. In addition, the mountains in this region are aligned northwest–southeast. The geomorphology of this region is a result of intense neotectonic activity, the direction of which is consistent with that of the tectonic lineaments in the region. Seismic activity occurs frequently in this region. This area is located in the northeastern Tibetan block, which experiences the most intense northeastward compression in the Tibetan block due to the northward

movement of the Indian Plate and the collision between the Eurasian Plate and the Indian Plate (Fig. 1a). The Tibetan block is one of the six intra-plate blocks within mainland China and is the block with the most intense seismic activity in China (Deng et al. 2014).

The tectonic zone at the Himalayan Plate boundary represents the southern boundary of the Tibetan block. The Tibetan block is composed of smaller blocks, including the Lhasa block, the Qiangtang block, the Bayan Har block, the East Kunlun–Qaidam block and the Qilian block (Zhang et al. 2003). Eighteen earthquakes with magnitudes ≥8.0 and more than 100 M7.0–7.9 earthquakes have occurred between the region's southern margin (boundary with the Himalayan Plate) and northern margin (Hexi Corridor) since records began (1900). All of the ≥M8.0 earthquakes occurred along the Himalayan Plate boundary, the boundary of the Tibetan fault block and the active tectonic zones on the boundaries of the secondary intra-plate fault blocks. These ≥M8.0 earthquakes have shallow epicenters (Deng et al. 2014).

The Yushu observation well is 105 m deep. The Quaternary sedimentary layer from the surface to a depth of 0.4 m consists mainly of silt mixed with a small amount of small gravel (Fig. 1b). The layer from 0.4 to 3.6 m consists mainly of angular gravel with skeletal particles arranged in a crisscross pattern. The gravel in this layer is mainly composed of moderately weathered angular granite and quartzite stones, which are generally 20–90 mm in size, filled with sand and a large amount of argillaceous matter. The layer below 3.6 m consists of fine grained and massive hard granite. The casing inside the well extends to a depth of 11 m (Fig. 2a). Fissure water was observed at 81 m below the surface (He et al. 2012; Yang et al. 2016), and the water temperature gradient shows a significant decrease below that depth (Fig. 2b).

Both the YSWT and the water level in the Yushu well (YSWL) are currently recorded (Fig. 3a). Observation of the YSWT began in October 2007. A SZW-1A digital quartz thermometer probe with a resolution of 0.0001 °C, sampling frequency of 1 min

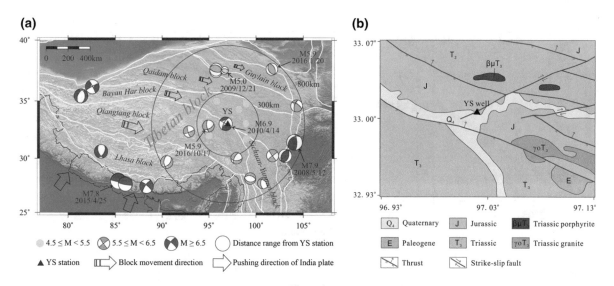

Figure 1

Tectonic (**a**) and geological (**b**) background of Yushu station, and the distribution of significant earthquakes in Tibetan block from October 1, 2007 to January 15, 2017. The *white lines* in tectonic map represent the boundary of block

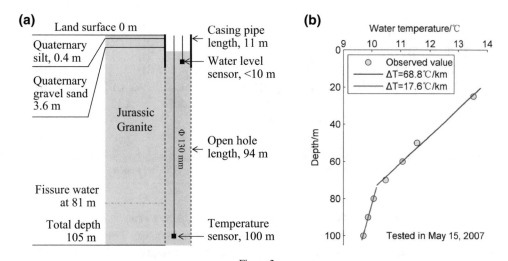

Figure 2

Structure of the Yushu well and lithology of the surrounding rocks (**a**), and the relationship between water temperature and depth in the well (**b**)

and pressure resistance greater than 10 MPa is located at a depth of 100 m to monitor the YSWT. The rock layer at this depth consists of Mesozoic Jurassic granite (Fig. 2a). Observation of the YSWL began in May 2014, by using a SWY-II water-level meter. A differential pressure sensor with a resolution of 0.001 m and a sampling rate of 1 per second is located at a depth of 10 m (Wang et al. 2016a; Yang et al. 2016).

## 3. Preseismic Changes in Temperature

### 3.1. Dynamic Changes in YSWT

Nearly continuous data have been obtained since the YSWT observations began in October 2007. The YSWT data from February 6–April 9, 2009 and June 21, 2012–July 19, 2013 are missing because of instrument failures during these two periods. The

41

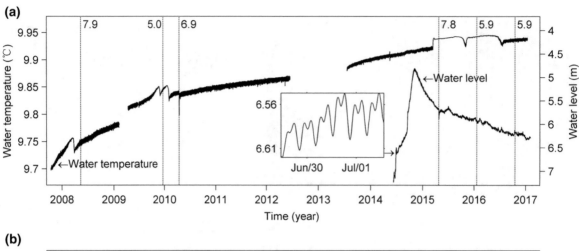

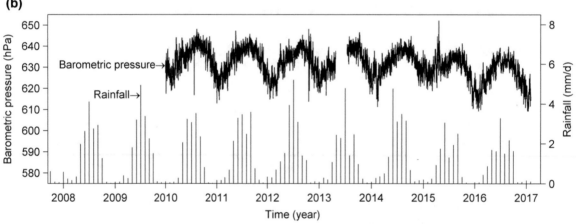

Figure 3
YSWT and YSWL (here, water level means the distance between the well head and the water surface inside the well) obtained at the Yushu well (**a**), and the barometric pressure and rainfall around Yushu station (**b**)

figure depicting the long-term dynamic changes in the YSWT from August 2007 to January 2017 shows that the YSWT increased year by year, and the changes in water temperature are almost insensitive to rainfall and barometric pressure (Fig. 3b).

The changes in the YSWT were characterized by an increasing trend over multiple years. However, the YSWT also underwent six significant abnormal changes during the observation period, and an earthquake occurred after each abnormal change (Fig. 3a). Except for the abnormal increase in the YSWT before the M7.8 earthquake in Pokhara, Nepal on April 25, 2015, the YSWT underwent "V"-shaped

abnormal changes (an initial decrease followed by a subsequent increase) before all of the earthquakes (the M7.9 earthquake in Wenchuan, Sichuan on May 12, 2008, the M5.0 earthquake in Delingha, Qinghai on December 21, 2009, the M6.9 earthquake in Yushu, Qinghai on April 14, 2010, the M5.9 earthquake in Menyuan, Qinghai on January 21, 2016 and the M5.9 earthquake in Zaduo, Qinghai on October 17, 2016). Among the six earthquakes, five of them occurred after the YSWT had undergone a "V"-shaped abnormal change. Therefore, abnormal changes in the YSWT may be related to seismic activity.

## 3.2. Relationship Between the Characteristics of YSWT Anomalies and Earthquakes

To further analyze the characteristics of the abnormal changes in the YSWT before the six earthquakes and explore the relationship between YSWT anomalies and earthquakes, the curves showing the abnormal dynamic changes in the YSWT before the earthquakes are plotted on one graph for comparison through time-translation operations (Fig. 4). The details of six anomalies are shown in Table 1, such as earthquake magnitude ($M$), epicentral distance ($S$), seismic energy density ($e$), extent of temperature change ($\Delta T$) and duration ($D$).

We defined the early period duration ($t < 0$) as the time when the YSWT starts to decrease (for V-type change) or increase (for step-rise change) to the time when the YSWT reaches the minimum or maximum value ($t = 0$), and the later period duration ($t > 0$) as the time from the minimum value or the maximum value of YSWT to the time of the earthquake. As shown in Fig. 5a, the greater the earthquake magnitude, the shorter the duration (early period or later period), except for Delingha M5.0 earthquake which deviates this relationship. Similar relations can be found for duration: the greater the magnitude of the earthquake, the greater the magnitude of temperature change (Fig. 5b).

Overall, the amplitude and duration of a YSWT anomaly were somewhat linearly related to the magnitude of the following earthquake. However, this correlation was affected by the distance between the well and the earthquake. For example, the YSWT changed the most ($-2.94 \times 10^{-2}$ °C) before the M6.9 earthquake in Yushu, Qinghai on April 14, 2010, whereas the well-epicenter distance was the smallest for this earthquake (only 22 km), and a notable co-seismic response also occurred (curve ③ in Fig. 4). While the two M5.9 earthquakes in Qinghai in 2016 had the same magnitude, the YSWT changed by different amplitudes before these two earthquakes; the smaller the well-hypocenter distance, the greater the change in the YSWT. This pattern suggests that the magnitude of a change in the YSWT may be indicative of

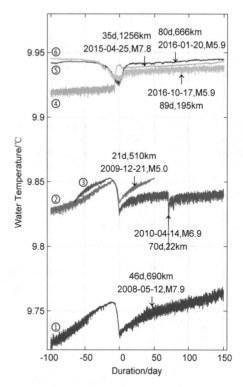

Figure 4

Curves showing the six significant abnormal changes in the YSWT, where curve ① shows the data for December 18, 2007–August 24, 2008, which correspond to the M7.9 earthquake in Wenchuan, Sichuan; curve ② shows the data for August 22, 2009–January 19, 2010, which correspond to the M5.0 earthquake in Delingha, Qinghai; curve ③ shows the data for November 30, 2009–July 3, 2010, which correspond to the M6.9 earthquake in Yushu, Qinghai; curve ④ shows the data for July 24, 2015–March 30, 2016, which correspond to the M7.8 earthquake in Pokhara, Nepal; curve ⑤ shows the data for April 11, 2016–December 17, 2016, which correspond to the M5.9 earthquake in Menyuan, Qinghai; curve ⑥ shows the data for January 1, 2008–July 31, 2008, which correspond to the M5.9 earthquake in Zaduo, Qinghai

the location of a future earthquake; i.e., when the YSWT changes by a relatively large amount, a future earthquake may occur relatively near the Yushu well, and vice versa.

As shown in Fig. 1a, most of the earthquakes that occurred after the YSWT experienced an abnormal change took place at the boundaries of the Tibetan fault block or its sub-blocks. The M6.9 Wenchuan earthquake and the M6.9 Yushu earthquake occurred on the active fault zone at the Bayan Har block boundary (Deng et al. 2010), and the M7.8 Pokhara

Table 1

*Large earthquakes that occurred after abnormal changes in the YSWT*

| Date (YYYY-MM-DD) | Latitude/ N° | Longitude/ E° | Location | Magnitude M | Epicentral distance/km | Energy density/Jm$^{-3}$ | Extent of change/ ($10^{-2}$ °C) | Duration/d | |
|---|---|---|---|---|---|---|---|---|---|
| | | | | | | | | Early period | Later period |
| 2008-05-12 | 31.44 | 104.10 | Wenchuan, China | 7.9 | 690 | $4.4 \times 10^{-2}$ | −2.88 | −11 | 46 |
| 2009-12-21 | 37.57 | 96.35 | Delingha, China | 5.0 | 510 | $6.7 \times 10^{-6}$ | −1.87 | −8 | 21 |
| 2010-04-14 | 33.05 | 96.79 | Yushu, China | 6.9 | 22 | $5.3 \times 10^{1}$ | −2.94 | −13 | 70 |
| 2015-04-25 | 27.91 | 85.33 | Pokhara, Nepal | 7.8 | 1256 | $5.2 \times 10^{-3}$ | 2.24 | −8 | 43 |
| 2016-01-20 | 37.68 | 101.62 | Menyuan, China | 5.9 | 666 | $6.1 \times 10^{-5}$ | −1.3 | −47 | 80 |
| 2016-10-17 | 32.81 | 94.93 | Zaduo, China | 5.9 | 195 | $2.5 \times 10^{-3}$ | −1.5 | −35 | 89 |

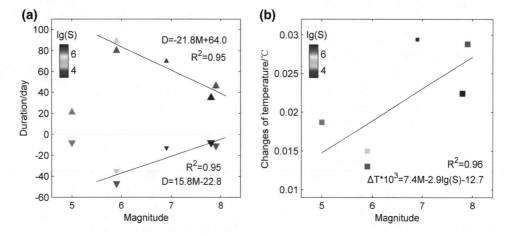

Figure 5

Relationship between M, D and S (**a**), and the relationship between *M*, *S* and Δ*T* (**b**), where *M*, *D*, *S* and Δ*T* mean earthquake magnitude, duration, epicentral distances [with the size and color means lg(*S*)] and the changes of water temperature, respectively

earthquake occurred in the Himalayan tectonic zone. These earthquakes were tectonically related to the Tibetan block (Bilham et al. 2001). Furthermore, both the M5.9 Menyuan earthquake and the M5.9 Zaduo earthquake occurred along the northeastern margin of the Tibetan Plateau. In addition, except for the M6.9 Yushu earthquake which produced maximum seismic energy density of 53 J/m³, the earthquakes following an YSWT anomaly all occurred on the thrust fault.

The above analysis shows that the YSWT responds well to earthquakes occurred in the Tibetan Plateau and its margins and that the preseismic anomalies in the YSWT are likely correlated with the regional tectonic structure. As shown in Fig. 1, the Yushu well was drilled into a fault zone and near the intersection of two fault zones. This may be why it is especially sensitive to premonitory groundwater variations. Under the pushing motion of the Indian plate, the sub-blocks of the Tibetan block keep moving and the preseismic strains may be concentrated near the faults at the block boundaries. Wells drilled into the fault zone at block boundaries, such as Yushu well, will respond to the strain concentration, and the next earthquake may occur on one of the faults. The larger the magnitude of the earthquake, the more energy was accumulated and the greater the extent of abnormal changes. The faster the strain accumulated, the shorter the duration of the anomaly (Fig. 5).

## 4. Tests Using the Molchan Error Diagram

Based on the analysis described above, the YSWT had anomalies before large earthquakes in the Tibetan Plateau and its margins. However, the YSWT did not change before every earthquake. Therefore, the effectiveness of using YSWT data to predict earthquakes is quantitatively examined using the Molchan error diagram method. As one of the six statistical test methods currently used by the Collaboratory for the Study of Earthquake Predictability (http://www.cseptesting.org), the Molchan error diagram method can objectively and scientifically predict and evaluate earthquakes, predict the timing of earthquakes in a given area, and provide corresponding probability interpretations (Jordan 2006). The Molchan error diagram method has been widely used in statistical tests and evaluations of the effectiveness of routine deterministic and probabilistic predictions (Zechar and Jordan 2008).

The area between 25°N and 40°N and 75°E and 110°E is selected for analysis. The earthquake catalog and focal mechanism solutions used in the analysis are extracted from the US Geological Survey Earthquake Catalog (http://earthquake.usgs.gov/earthquakes/search/), the aftershocks are removed, and the time is adjusted to Beijing Standard Time. A total of 36 earthquakes (16 M4.5–5.5 earthquakes within 300 km of the Yushu well, 11 M5.5–6.5 earthquakes within 800 km of the Yushu well and 9 ≥ M6.5 earthquakes in the Tibetan Plateau) are selected. Figure 1 shows the distribution of the epicenters of the selected earthquakes.

### 4.1. Molchan Error Diagram

The Molchan error diagram method mainly examines the difference between the predicted results and the observed data of the target earthquake (Molchan 1990). The method not only visually reflects the prediction performance and evaluates the observed data but also can quantitatively analyze

the outliers and obtain the outlier identification indices that correspond to the optimum threshold. The Molchan error diagram method calculates the following variables:

1. Hit rate ($H$): ratio of the number of predicted earthquakes to the total number of actual earthquakes.
2. Miss rate ($v$): ratio of the number of unpredicted earthquakes that actually took place to the total number of actual earthquakes.
3. Fraction of space–time occupied by the alarm ($\tau$): ratio of the spatiotemporal range within which outliers are extracted using different thresholds to the total spatiotemporal range.

Only the fraction of time occupied by the alarm needs to be considered when using the Molchan error diagram method to examine the effectiveness of using observed YSWT data to predict earthquakes. By continuously decreasing the "alarm" threshold for earthquake prediction, the fraction of time occupied by the alarm ($\tau$) and the corresponding miss rate ($v$) are calculated, based on which a $\tau$–$v$ curve is plotted in a Molchan error diagram. The area enclosed by the $\tau$–$v$ curve and the boundary of the diagram indicates the prediction performance of the $\tau$–$v$ curve compared to a stochastic prediction. The smaller the area is, the better the prediction performance is. However, it is also necessary to examine the probability gain (Gain) and significance level ($\alpha$). In the Molchan error diagram method, they are defined as follows (Molchan 1991; Zechar and Jordan 2008):

$$\text{Gain} = \frac{H}{\tau} = \frac{1 - v}{\tau}$$

$$\alpha = \sum_{i=h}^{N} \left[ \frac{N!}{i!(N-i)!} \tau^i (1 - \tau)^{N-i} \right]$$

where $N$ is the total number of actual earthquakes and $h$ is the number of predicted earthquakes.

The greater the probability gain, the better the prediction performance. A $\tau$–$v$ curve that is close to the straight line of the probability Gain = 1 indicates that the prediction results are not statistically significant. The actual test process is as follows. After a threshold is determined, the data that exceed this threshold are deemed outliers. An earthquake is correctly predicted if it occurs in the time period during which an outlier occurs as well as in its effective prediction period. An earthquake is missed by the prediction if it occurs in a time period outside the time period during which an outlier occurs as well as its effective prediction period. The fraction of time occupied by the alarm is obtained by dividing the total length of all of the anomalous times and the effective prediction periods (with repeats removed) by the total length of time of the examined data.

## 4.2. Test Results for the YSWT

In Fig. 6, the blue curves show the original observed daily values of the YSWT, and the green curves depict the absolute values of the five-day differential values of the YSWT. The arrows show the 36 earthquakes; the pink arrows represent the <M5.5 earthquakes, and the red arrows represent the ≥M5.5 earthquakes. Over several years, the YSWT had a gradual increase; however, the YSWT underwent six significant abnormal changes during this period. These six changes are subjected to a 5-day differential transformation. The differential results show the changes in the high absolute values more prominently. An earthquake occurred after each high-value anomaly. However, not all of the earthquakes were preceded by an anomaly.

The data also show that a relatively large magnitude earthquake and several low magnitude earthquakes occurred after abnormal changes in the YSWT. This presents difficulty for the prediction of

possible earthquakes based on the analysis of abnormal changes in the YSWT.

Figure 7 shows the Molchan test results for the YSWT with the ≥M4.5 earthquakes and ≥M5.5 earthquakes shown in Fig. 1 (statistical time window length: 90 days). Figure 7a, b shows the τ–v curves and the probability gain contours for the ≥M4.5 earthquakes and the ≥M5.5 earthquakes, respectively [x-axis: fraction of time occupied by the alarm (τ); left y-axis: miss rate (v)]. The stochastic prediction line (probability gain = 1) separates the diagram into two parts: the lower left part and the upper right part. Each threshold determines a set of τ–v values. A τ–v curve is obtained by incrementally increasing the threshold. The area enclosed by the τ–v curve, the x-axis and the left y-axis indicates the prediction performance. The smaller the gray area is, the better the prediction performance is, and vice versa.

In Fig. 7a, b, point (1, 0) in the lower right corner indicates that all of the earthquakes are accurately predicted, but the fraction of time occupied by the alarm (false alarm rate) at this point is also the largest because it covers the entire time period of the data. The left y-axis represents the miss rate (v), and the right y-axis represents the hits (h). The higher the miss rate is, the lower the hit rate is. For example, point (0, 1) in the upper left corner indicates that all of the earthquakes are missed by the prediction (none of the earthquakes are accurately predicted), but the fraction of time occupied by the alarm at this point is the smallest, which is equivalent to no prediction being made.

A comparison of the gray areas enclosed by the τ–v curves, y-axes and x-axes in Fig. 7a, b shows that the test results of the YSWT obtained for the ≥M5.5 earthquakes has a smaller area than the results for the ≥M4.5 earthquakes, which indicates better prediction performance. In addition, when the ≥M5.5 earthquakes are selected, most of the predictions have a probability gain of approximately 2 and significance level between 1 and 5%. Therefore, for the same time window (90 days), using observed YSWT data is more effective at predicting ≥M5.5 earthquakes than ≥M4.5 earthquakes.

Figure 8 shows the effectiveness of using observed YSWT data to predict earthquakes with various time windows (1–365 days; statistically

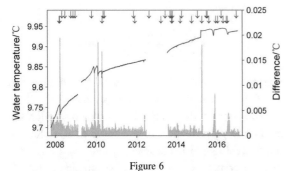

Figure 6

Daily values (*blue curves*) and five-day differential values (*green curves*) of the YSWT and the earthquakes in the regions surrounding the Yushu well (the *arrows* show the 36 earthquakes, the *pink arrows* represent the *M* < 5.5 earthquakes, and the *red arrows* represent the *M* ≥ 5.5 earthquakes)

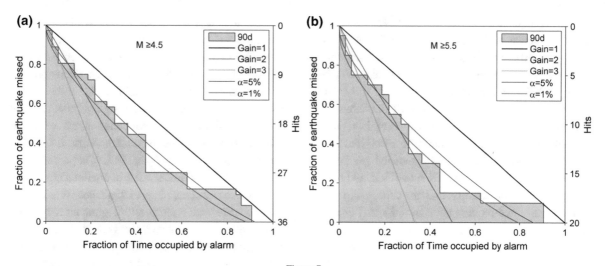

Figure 7
Molchan test results for the YSWT: **a** ≥M4.5 earthquakes; **b** ≥M5.5 earthquakes

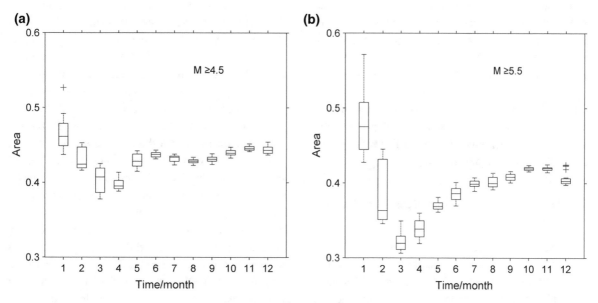

Figure 8
Examination of the effectiveness of using observed YSWT data to predict earthquakes with various time windows: **a** ≥M4.5 earthquakes; **b** ≥M5.5 earthquakes. In a *boxplot*, the *upper* and *lower edges* of the *box* represent the first and third quartile (Q1 and Q3), respectively, of the dataset. The *horizontal line* segment inside the *box* represents the median. There are another two line segments outside the *box* representing Q3 + 1.5 □ IQR and Q3 − 1.5 □ IQR (IQR is the interquartile range, i.e., Q3–Q1). These two segments correspond to the outlier truncation points, which constitute the so-called inner limits. Data points inside the inner limits are considered to represent normal values for this dataset, whereas those outside the inner limits are considered outliers (marked using *red* "+" *symbols*)

analyzed on a monthly basis). In Fig. 8, the *x*-axis represents time (12 months), and the y-axis represents the gray area in Fig. 7, which indicates the effectiveness of using observed YSWT data to predict earthquakes for various time windows (the smaller the area, the higher the effectiveness). Figure 8a shows the effectiveness of using observed YSWT data to predict ≥M4.5 earthquakes, and Fig. 8b shows the effectiveness of using observed YSWT data to predict ≥M5.5 earthquakes. For predicting

≥M4.5 earthquakes, most of the effectiveness values are in the range of 0.4–0.5 with insignificant differences for different time windows (1–12 months). However, in comparison, the effectiveness of predicting ≥M5.5 earthquakes varies significantly with the time window. The observed YSWT data are most effective for predicting earthquakes with a time window of approximately three months (smallest area). Therefore, the YSWT is somewhat predictive of ≥M5.5 earthquakes in the Tibetan block, and there is a significant probability that an earthquake will occur within three months after an abnormal change in the YSWT.

## 5. Discussion

By analyzing the significant anomalies in the YSWT that have occurred since observations began in October 2007, this study found that except for the increase in the YSWT before the ≥M7.8 earthquake in Pokhara, Nepal in April 2015, the abnormal changes before all of the other earthquakes exhibited similar "V"-shaped patterns. The quantitative results of the Molchan error diagram test show that the YSWT is somewhat predictive of ≥M5.5 earthquakes in the Tibetan block and its margins.

### 5.1. Analysis of the Mechanism of the Preearthquake YSWT Anomalies

Previous studies have investigated the mechanism of the changes in water temperatures in monitoring wells and proposed several mechanisms. The rock mechanics mechanism proposes that the changes of water temperature in a well are a direct result of the stress-induced deformation of the aquifer, as fresh fractures are generated before an earthquake (Hadley 1973; Fu 1988; Liu et al. 1989). Che et al. (1998) concluded that main factor affecting the water temperature in a well is the water flows with different temperature between the well and the aquifer, termed the hydro-thermodynamic mechanism. Orihara et al. (2014) assumed that expansion around the confined groundwater region produced paths for water, which caused the groundwater temperature to decrease due to inflow of colder water from other aquifers.

Che and Yu (1997) reports that not all earthquake precursors include information about the earthquake focus. Earthquake precursors can be classified as focus precursors and field precursors. Focus precursors are anomalies that originate from the focus, whereas field precursors refer to regional tectonic activity anomalies that are concomitant with the earthquake process. The tectonic activity in a region is somewhat hereditary. Therefore, it is possible that the YSWT will exhibit similar anomalous characteristics before different large earthquakes. In addition, based on the water temperature gradient when the Yushu well was completed, the groundwater temperature gradually decreases with increasing depth in this well; i.e., there is a negative gradient between the groundwater temperature and the depth (Wang et al. 2016a, b). A turning point is located in the well at a depth of approximately 80 m (Fig. 2b), below which the change in the water temperature gradually decreases. This indicates that the "V"-shaped abnormal change in the YSWT before a large earthquake might be caused by the convection of water with different temperature between the wellbore and the aquifer, and the convection may be related to the preseismic strain field accumulation near the fault.

Based on the hydro-thermodynamic mechanism, it is possible that the abnormal changes in the YSWT were caused by heat convection due to water movement, which was caused by the stress-induced deformation of the aquifer. As shown in Fig. 2, fissure water is present at a depth of approximately 80 m in the Yushu well, and the temperature probe is located at a depth of 100 m. The movement of the fault (e.g., slow slip) near a well will cause a water flow between the well and the aquifer. Due to the vertical and horizontal water temperature gradients, the flow of water between the well and the aquifer will result in heat exchange, which in turn results in an abnormal change in the water temperature. But there were no abnormal changes in YSWL while YSWT undergoing abnormal changes (Fig. 3a). So, the hypothesis about water flow does not seem to be supported by water-level changes.

The Nepal M7.8 earthquake in 2015 induced coseismic water-level change, but did not induce coseismic change in water temperature. There are two possible reasons for this difference, (1) water temperature reflects the transfer of heat while the water level reflects the transfer of pressure; (2) the probe of water temperature was placed in the bottom of well. When the seismic waves passed the well-aquifer system, pressure (or permeability) in aquifer could be changed and cause the water flow between well and aquifer at 81 m (Fig. 2). However, the volume of water flow might not enough to induce the heat changes in bottom of hole or the heat changes was too small to detect; thus, the water temperature would not respond to the seismic waves.

### 5.2. Water-Level Record of YSWT Anomalies

The YSWT underwent "V"-shaped abnormal changes that were marked by an initial decrease followed by a subsequent increase before multiple large earthquakes. However, the YSWT abnormally increased before the M7.8 earthquake in Pokhara, Nepal in 2015. This suggests that the preseismic response of the water temperature in a well also varies with the earthquake. The analysis described above reveals that changes in the water temperature in a well are related to the movement of water in the well–aquifer system.

Water-level observation in Yushu well began in 2014 (Fig. 3a). The YSWL shows a relatively good signature of the solid Earth tide, which indicates that the Yushu well is a well confined well-aquifer system. This provides a reference for studying the correlation between the abnormal changes in the YSWT and the tectonic activity in the surrounding regions. The tidal response of the water level in a well is closely related to the characteristics of the well-aquifer system (Hsieh et al. 1988), and the phase lag of the tidal response is influenced by changes in the permeability of the aquifer (Brodsky et al. 2003; Shi and Wang 2016; Wang et al. 2016a, b; Yan et al. 2016) and is thus closely correlated to the porosity and stress-induced deformation of the aquifer. In this study, the phase lag of the YSWL is extracted using the Venedikov harmonic analysis method (Venedikov et al. 2003) (Fig. 9). The results show that the phase lag of the YSWL first increased and then decreased before the 2015 M7.8 Pokhara earthquake and that there was a significant difference in the phase lag of the YSWL before and after this earthquake. Furthermore, the phase lag of the YSWL decreased slowly after the earthquake until the period before the October 2016 M5.9 Zaduo earthquake, when the phase lag of the YSWL exhibited a similar pattern (an initial increase followed by a subsequent decrease) to that before the M7.8 Pokhara earthquake.

Possible causes for changes in the phase lag of the water level (tide) are the change of recharge or permeability of the Yushu well-aquifer system due to the stain accumulation before and after the 2015 M7.8 Pokhara earthquake, which was manifested by

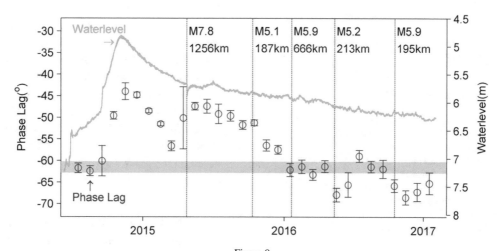

Figure 9
Phase lag of earth tide (*blue error bar*) and the water level (*green line*) in the Yushu well. The *horizontal gray line* indicates the general state without the effect of earthquake

an initial increase in the phase lag of the YSWL followed by a decrease. After the earthquake, the intra-plate tectonic activity in the Tibetan Plateau slowly recovered and adjusted; this was manifested by a slow decrease in the phase lag of the YSWL, which, after the M5.9 Menyuan earthquake, had essentially recovered to the level before the M7.8 Pokhara earthquake. While several M5–6 earthquakes occurred during this period, they only had a minor impact and did not significantly affect the overall adjustment process. The change in the phase lag of the YSWL before the M5.9 Zaduo earthquake, which occurred 195 km from the Yushu well, had a pattern similar to that before the M7.8 Pokhara earthquake and was likely a result of the deformation of the aquifer by the tectonic activity during the earthquake in Zaduo, Qinghai.

The analysis of the phase lag of the water level (tide) further reveals that the Yushu observation well is located at a sensitive location in the tectonic structure (with tide factor 2–5 mm/$10^{-9}$) and that the YSWL and YSWT respond relatively well to earthquake processes in the nearby region as well as in the Tibetan block and its margins. Thus, the YSWT underwent "V"-shaped abnormal changes

Table 2

*List of features that need to be documented for YSWT*

| | Features | Description in paper |
|---|---|---|
| 1 | Depth of well | Shown Fig. 2a |
| 2 | Rainfall | Shown Fig. 3b |
| 3 | Barometric pressure | Shown Fig. 3b |
| 4 | Pumping and injection in the vicinity | There are no pumping and injection within 50 km |
| 5 | Entire observation record | Shown Fig. 3a |
| 6 | Measurement technique | Shown Sect. 2 |
| 7 | Sampling interval | Minutes for water temperature, Seconds for water level |
| 8 | Response to earth tides | Shown Figs. 3a and 9 |
| 9 | Co-seismic and post seismic response to the earthquake | Shown Figs. 3a and 4 |
| 10 | Earthquake magnitude, azimuth, distance, depth and focal mechanism | Shown Fig. 1a and Table 1 |
| 11 | Time, location and magnitude of any foreshocks | None |
| 12 | Raw water-level data | Shown Fig. 3a |
| 13 | Description of other wells in the area that did not document the anomaly | There are no other water temperature observation wells within 100 km |

before several earthquakes. While the original curve of the YSWL exhibits no abnormal changes, the phase lag of the tide in the Yushu well exhibits clear anomalies. These anomalies may reflect the pushing action of the Tibetan plate, the adjustments of the regional stress, and the development and occurrence of earthquakes.

## 6. Conclusions

By analyzing the correlation between several significant abnormal changes in the YSWT and earthquakes in the surrounding region as well as in the Tibetan block and its margins, this study found that the YSWT is somewhat predictive of earthquakes in these regions. The effectiveness of using observed YSWT data to predict earthquakes was quantitatively examined using the Molchan error diagram method, and the optimum time period for predicting earthquakes based on YSWT was obtained. The following conclusions are drawn:

1. An earthquake occurred in the Tibetan block or its margins after each of the significant abnormal changes in the YSWT. These abnormal changes in the YSWT were marked by similar "V"-shaped patterns. The extent and duration of the abnormal change in the YSWT were linearly related to the magnitude of the earthquake; i.e., the greater the change in the YSWT, the higher the magnitude, and the shorter the duration, the smaller the earthquake.

2. The results of the Molchan error diagram tests show that abnormal changes in YSWT are somewhat predictive of ≥M5.5 earthquakes within 800 km of the Yushu well and of ≥M6.5 earthquakes in the Tibetan block and its margins. The most likely time period for an earthquake to occur based on the YSWT is within approximately three months after the occurrence of an YSWT anomaly.

Groundwater temperature anomalies associated with earthquakes are well known, but the cause of the anomalies remains unclear. The mechanism of changes in the water temperature in a well is relatively complex and may involve many factors, such as the hydrogeological conditions of the well, the geothermal conditions, and the regional tectonic setting. The conclusions about the correlation between abnormal changes in the YSWT and earthquakes are mainly based on the statistical analysis of the observed phenomena. Although we attempted to explain the abnormal changes of temperature, we are not able to identify the exact causative mechanism. For example, five anomalies listed in Table 1 showed V-shape patterns and one anomaly (Nepal earthquake in 2015) showed a different pattern: different patterns suggest different mechanisms, but we can not identify the difference.

Further study is needed in order to obtain a more quantitative and physical explanation for the abnormal changes in the water temperature. A more definitive analysis of the mechanism requires the accumulation of more data. As noted by Roeloffs (1998), poor documentation is the major impediment to using and interpreting water-level data. The same should be true for temperature data. Our study documented all the information proposed by Roeloffs (1998) (Table 2) and thus we hope for our study will have important implications for earthquake precursor studies.

## Acknowledgements

We would like to thank the Guest Editor Michael Manga and the reviewers for their constructive comments which greatly improve the manuscript. The authors wish to thank the China Earthquake Network Center and the Qinghai Provincial Seismological Bureau for providing the water temperature and level data in the Yushu well. This study was financially supported by the National Natural Science Foundation of China (41502239, U1602233) and the Monitoring, Prediction and Research-combined Project of the China Earthquake Administration under (163102).

## References

Asteriadi, G., & Liverato, E. (1989). Pre-seismic responses of underground water level and temperature concerning a 4.8 magnitude earthquake in Greece on October 20. *Tectonophysics, 170,* 165–169.

Bilham, R., Vinod, K., & Molnar, P. (2001). Himalayan seismic hazard. *Science, 293,* 1442–1444.

Brodsky, E. E., Roeloffs, E., Woodcock, D., Gall, I., & Manga, M. (2003). A mechanism for sustained groundwater pressure changes induced by distant earthquakes. *Journal of Geophysical Research: Solid Earth, 108*(B8), 2390.

Caine, J. S., & Minor, S. A. (2009). Structural and geochemical characteristics of faulted sediments and inferences on the role of water in deformation, Rio Grande Rift, New Mexico. *Geological Society of America Bulletin, 121*(9–10), 1325–1340.

Che, Y. T., Liu, W. Z., & Yu, J. Z. (1998). The relationship between crustal fluids and major earthquakes and its implications for earthquake prediction. *Seismology and Geology (in Chinese), 20*(4), 431–435.

Che, Y. T., & Yu, J. Z. (1997). The focal precursors, field precursors and remote precursors of ground fluid before strong earthquakes and their significance in the earthquake prediction. *Earthquake (in Chinese), 17*(3), 283–289.

Deng, Q. D., Cheng, S. P., Ma, J., et al. (2014). Seismic activities and earthquake potential in the Tibetan Plateau. *Chinese J Geophys. (in Chinese), 57*(7), 2025–2042. doi:10.6038/cjg20140701.

Deng, Q. D., Gao, X., Chen, G. H., & Yang, H. (2010). Recent tectonic activity of Bayankala fault-block and the Kunlun Wenchuan earthquakes series of the Tibetan Plateau. *Earth Science Frontiers (in Chinese), 17*(5), 163–178.

Du, L. T. (2007). Concept renewal from solid earth science to fluid one. *Progress in Geophysics (in Chinese), 22*(4), 1220–1224.

Faulkner, D. R., & Rutter, E. H. (2003). The effect of temperature, the nature of the pore fluid, and subyield differential stress on the permeability of phyllosilicate rich fault gouge. *Journal of Geophysical Research: Solid Earth, 108*(B5), 2227. doi:10.1029/2001JB001581.

Fu, Z. Z. (1988). Premonition and dynamic observation of terrestrial heat. *Corpus of Crust Structure andStress (in Chinese), 1,* 1–7.

Hadley, K. (1973). Laboratory investigation of dilatancy and motion on fault surfaces at low confining pressures. In R. L. Kovach & A. Nur (Eds.), *Proceedings of the conference on tectonic problems of the san andreas fault system* (Vol. XIII). Stanford: Stanford University Publications, Geological Sciences.

He, A. H., Zhao, G., & Liu, C. L. (2012). The anomaly characteristics before Wenchuan earthquake and Yushu earthquake in Qinghai Yushu and Delingha geothermal observation wells. *Chinese Journal of Geophysical (in Chinese), 55*(4), 1261–1268. doi:10.6038/j.issn.0001-5733.2012.04.021.

Hsieh, P. A., Bredehoeft, J. D., & Rojstaczer, S. A. (1988). Response of well aquifer systems to Earth tides: Problem revisited. *Water Resources Research, 24*(3), 468–472.

Johnston, M. J. S., Hill, D. P., Linde, A. T., Langbein, J., & Bilham, R. (1995). Transient deformation during triggered seismicity from the 28 June 1992 Mw = 7.3 Landers earthquake at Long Valley volcanic caldera, California. *Bulletin of the Seismological Society of America, 85,* 787–795.

Jordan, T. H. (2006). Earthquake predictability, brick by brick. *Seismological Research Letters, 77*(1), 3–6. doi:10.1007/s10950-008-9147-6.

Kitagawa, Y., Koizumi, N., & Tsuskutta, T. (1996). Comparison of post-seismic groundwater temperature changes with earthquake-induced volumetric strain release: Yundani hot spring, Japan. *Geophysical Research Letters, 23*(22), 3147–3150.

Liu, L. B., Roeloffs, E., & Zheng, X. Y. (1989). Seismically induced water level fluctuations in the Wali well, Beijing, China. *Journal of Geophysical Research: Solid Earth, 94*(B7), 9453–9462, 1978–2012.

Manga, M., Beresnev, I., Brodsky, E. E., Elkhoury, J. E., Elsworth, D., Ingebritsen, S. E., et al. (2012). Changes in permeability caused by transient stresses: Field observations, experiments, and mechanisms. *Review of Geophysics, 50,* RG2004.

Mogi, K., Mochizuki, H., & Kurokawa, Y. (1989). Temperature changes in an artesian spring at Usami in the Izu Peninsula (Japan) and their relation to earthquakes. *Tectonophysics, 159*(1–2), 95–108.

Molchan, G. M. (1990). Strategies in strong earthquake prediction. *Physics of the Earth and Planetary Interiors, 61*(1–2), 84–98.

Molchan, G. M. (1991). Structure of optimal strategies of earthquake prediction. *Tectonophysics, 193,* 267–276.

Montgomery, D. R., & Manga, M. (2003). Streamflow and water well responses to earthquakes. *Science, 300*(5628), 2047–2049.

Orihara, Y., Kamogawa, M., & Nagao, T. (2014). Preseismic Changes of the Level and Temperature of Confined Groundwater related to the 2011 Tohoku Earthquake. *Scientific reports, 4.* doi:10.1038/srep06907

Roeloffs, E. A. (1998). Hydrologic precursors to earthquakes: A review. *Pure and Applied Geophysics, 126,* 177–209.

Sadovsky, M. A., Nersesov, I. L., Nigmatullaev, S. K., Latynina, L. A., Lukk, A. A., Semenov, A. N., Simbireva, I, G., Ulomov, V. I. (1972). The processes preceding strong earthquakes in some regions of middle Asia. *Tectonophysics, 14(3),* 295–307. doi: 10.1016/0040-1951(72)90078-9

Shi, Z., & Wang, G. (2014). Hydrological response to multiple large distant earthquakes in the Mile well, China. *Journal of Geophysical Research: Earth Surface, 119,* 2448–2459.

Shi, Z., & Wang, G. (2016). Aquifers switched from confined to semiconfined by earthquakes. *Geophysical Research Letters, 43,* 11,166–111,172.

Shi, Z., Wang, G., Manga, M., & Wang, C. Y. (2015). Mechanism of co-seismic water level change following four great earthquakes-insights from co-seismic responses throughout the Chinese mainland. *Earth and Planetary Science Letters, 430,* 66–74.

Shimamura, H., Ino, M., Hikawa, H., & Iwasaki, T. (1984). Groundwater microtemperature in earthquake regions. *Pure and Applied Geophysics, 122*(6), 933–946.

Simpson, D. W. (1986). Triggered earthquakes. *Annual Review of Earth and Planetary Science, 14*(1), 21–42.

Sun, X., & Liu, Y. (2012). Changes in groundwater level and temperature in induced by distant earthquakes. *Geosciences Journal, 16*(3), 327–337.

Venedikov, A. P., Arnoso, J., & Vieira, R. (2003). VAV: A program for tidal data processing. *Computers & Geosciences, 29*(4), 487–502. doi:10.1016/S0098-3004(03)00019-0.

Wakita, H. (1975). Water wells as possible indicators of tectonic strain. *Science, 189*(4202), 553–555.

Wang, C. Y., Liao, X., Wang, L. P., Wang, C. H., & Manga, M. (2016a). Large earthquakes create vertical permeability by breaching aquitards. *Water Resources Research, 52*(8), 5923–5937.

Wang, B., Ma, Y. C., & Ma, Y. H. (2016b). Variation of water temperature in Yushu well and its correlation with the big earthquakes occurred in Qinghai-Tibet block. *Earthquake Research in China, 32*(3), 461–468.

Wang, C. Y., Manga, M., Wang, C. H., & Chen, C. H. (2012). Transient change in groundwater temperature after earthquakes. *Geology, 40,* 119–122.

Wang, C. Y., Wang, L. P., Manga, M., Wang, C. H., & Chen, C. H. (2013). Basin-scale transport of heat and fluid induced by earthquakes. *Geophysical Research Letters, 40,* 1–5. doi:10.1002/grl.50738.

Yan, R., Wang, G., & Shi, Z. (2016). Sensitivity of hydraulic properties to dynamic strain within a fault damage zone. *Journal of Hydrology, 543,* 721–728.

Yang, X. F., Sun, L., Sun, C. L., & Li, Y. L. (2016). Relation between the water temperature anomaly of Yushu well and Nepal Ms8.1 earthquake. *Plateau Earthquake Research (in Chinese), 28*(2), 12–15.

Zechar, J. D., & Jordan, T. H. (2008). Testing alarm-based earthquake predictions. *Geophysical Journal International, 172,* 715–724.

Zhang, P. Z., Deng, Q. D., Zhang, G. M., et al. (2003). Strong earthquakes and active blocks in mainland China. *Chinese Science: Series D, 33*(Suppl.), 12–20. **(in Chinese)**.

(Received February 13, 2017, revised May 18, 2017, accepted May 19, 2017)

Pure Appl. Geophys.
© 2017 Springer International Publishing AG
DOI 10.1007/s00024-017-1710-z

**Pure and Applied Geophysics**

CrossMark

# Hydrological Changes Induced by Distant Earthquakes at the Lujiang Well in Anhui, China

YUCHUAN MA,[1,2] GUANGCAI WANG,[1] and YUECHAO TAO[3]

*Abstract*—The Lujiang well, a 63 °C artesian well, recorded sustained hydrological changes following the 1999 Chi–Chi $M_w$ 7.6, the 2008 Wenchuan $M_w$ 7.9, and the 2011 Tohoku $M_w$ 9.0 earthquakes, including rises in the water radon concentration, water pressure, discharge and water level, and drops in the water temperature. These hydrological changes are synchronous and have similar amplitudes. The permeability inferred through the tidal response of water level showed insignificant change after the three earthquakes. We attribute the observed hydrological changes to the increase in the vertical recharge on the basis that the water radon concentration of the Lujiang well increased following the increase of recharge to the well; significant vertical flow exists in the well-aquifer system; the well has a lower water radon concentration and a higher water temperature than its adjacent wells with different aquifers.

**Key words:** Groundwater radon concentration, water pressure, discharge, water level, water temperature, coseismic responses, permeability enhancement.

## 1. Introduction

It is well known that earthquakes can induce a variety of hydrological changes (Wang and Manga 2010a). Examples include changes in the water level or pressure (Cooper et al. 1965; Liu et al. 1989; Roeloffs 1998; King et al. 1999; Wang et al. 2009; Geballe et al. 2011; Weingarten and Ge 2014; Sun et al. 2015), water temperature (Shi et al. 2007; Wang et al. 2012, 2013; Shi and Wang 2014), and chemical composition (Wakita et al. 1989; Claesson et al. 2004; Huang et al. 2004; Ren et al. 2012; Skelton

**Electronic supplementary material** The online version of this article (https://doi.org/10.1007/s00024-017-1710-z) contains supplementary material, which is available to authorized users.

[1] School of Water Resources and Environment, China University of Geosciences, Beijing, China. E-mail: mayuchuan@seis.ac.cn
[2] China Earthquake Networks Center, Beijing, China.
[3] Anhui Earthquake Administration, Hefei, China.

et al. 2014). Among these, water level or pressure changes are widely reported and well studied. One plausible mechanism for sustained water level changes is earthquake-enhanced permeability (Mogi et al. 1989; Rojstaczer et al. 1995; Brodsky et al. 2003; Wang et al. 2004; Manga et al. 2012). This mechanism is supported by the permeability enhancement in the horizontal direction revealed through the analysis of the responses of well water level to Earth tides (Elkhoury et al. 2006; Xue et al. 2013; Lai et al. 2016). In addition, large earthquakes can switch confined aquifers to semiconfined ones (Shi and Wang 2016) or even create vertical permeability by breaching aquitards (Liao et al. 2015; Wang et al. 2016), suggesting that the permeability enhancement in the vertical direction can also change water levels. The permeability in the vertical direction can also be revealed through the analysis of the responses of well water level to Earth tides (Allègre et al. 2016; Xue et al. 2016). However, the responses of well water level to Earth tides do not change at some wells (Yan et al. 2014; Shi et al. 2015; Ma and Huang 2017), and the mechanisms responsible for the sustained water level changes at these wells have remained uncertain.

Here we report hydrological changes of different parameters following three distant earthquakes at a hot artesian well. We use the tidal response of water level to show that the permeability enhancement cannot explain the hydrological changes. We then attribute these changes to the increase in the vertical recharge, as supported by the water level, water temperature, and water radon concentration data.

## 2. Observations

The Lujiang well (31°20′N, 117°05′E, elevation 42.5 m) is located in a geothermal field at the

Tangchi Town, Lujiang County. The well site is located where the Tancheng–Lujiang fault, the Meishan–Longhekou fault, and the Qingsha–Xiaotian fault intersect (Fig. 1). The Lujiang well was drilled in November 1973 with a depth of 327.05 m. The well is artesian and its water temperature is 63 °C. Its structure, water temperatures at different depths and lithologic log are provided in Fig. 2. The core samples from 107.81 to 111.7 m and 291.8–306 m are shattered. The well has three aquifers between 90 and 185 m, and one aquifer between 290 and 300 m. The pumping test conducted when the Lujiang well was constructed indicates that the section between 42.83 and 105.25 m has the biggest storage coefficient ($1.46 \ \mathrm{Ls}^{-1} \ \mathrm{m}^{-1}$) and the biggest hydraulic conductivity (3.096 m/day). According to the chemical analyses of the well water sampled at several different times, the hydrochemical type of the well water is sodium sulfate, and the total dissolved solids, pH, and hardness are approximately 1.15 g/L, 8.3, and 5°dH, respectively (Anhui Earthquake Administration 2004; Bo et al. 2007).

The groundwater of the Lujiang well has been observed by the staff at the Lujiang station for seismological research since 1986. The station is a class-I groundwater observation station, the highest class of the China Earthquake Administration (CEA). The observation parameters include the water radon concentration, water pressure, discharge, water level and water temperature. Some basic information regarding the parameters is given in Table 1. The SZW-1A in Table 1 records the water temperature automatically through a quartz thermometer. The RGP-1 and the LN-3A in Table 1 record the water level automatically through pressure transducers. Before January 2000, the water temperature measured by the SZW-1A and the water level measured by the RGP-1 were printed on paper media. Thereafter, the water temperature measured by the SZW-1A and the water level measured by the LN-3A have been outputted to digital media.

The method of measuring the water radon concentration involves the following procedures. A volume of ∼ 100 ml of water is sampled into a glass diffuser under negative pressure at ∼ 8 am

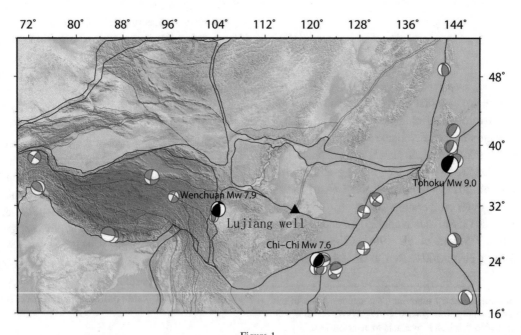

Figure 1
Locations of the Lujiang well (triangle) and some of the earthquakes ("beach balls") that caused hydrological changes. The "beach balls" are focal mechanisms of the earthquakes. Earthquakes highlighted in dark and with texts are the 1999 Chi–Chi $M_{\mathrm{w}}$ 7.6, the 2008 Wenchuan $M_{\mathrm{w}}$ 7.9 and the 2011 Tohoku $M_{\mathrm{w}}$ 9.0 earthquakes. Tick lines are boundaries of active tectonic block regions, and thin lines are boundaries of active tectonic blocks (Zhang et al. 2003)

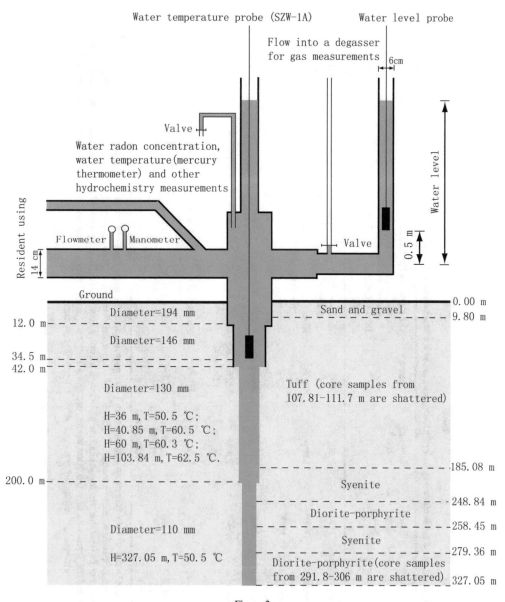

Figure 2

Illustration showing the structure, lithologic log, temperature and groundwater observations of the Lujiang well. This illustration is from Zhang (1991) and Anhui Earthquake Administration (2004). H and T stand for the depth and the water temperature, respectively. The water temperature data that were measured when the well was drilled are from Bo et al. (2007)

every morning. The naturally cooled water is bubbled to capture gaseous phases. The gaseous phases are transmitted into a scintillation chamber. The inner surface of the scintillation chamber is covered with a $\sim 80$ mg/cm$^2$ thick ZnS(Ag). In the scintillation chamber, the alpha particle incidents, originating from the decay of Rn$^{222}$, make the ZnS(Ag) scintillated and generate photons. The photons are converted into pulse voltages through a photomultiplier tube. The pulse voltages are amplified by a preamplifier. The pulse frequency is counted by a scaling block an hour after the gaseous phases are transmitted. The pulse frequency ($N$) is converted into the radon concentration ($C$) by the formula:

Table 1

*Instrument information and the sustained hydrological responses to three distant earthquakes of the Lujiang well*

| Observation parameter | Recorder | Manufacturer | Initial observation time | Sampling interval | Accuracy | The type, amplitude, and recovery time of hydrological changes after earthquakes | | |
|---|---|---|---|---|---|---|---|---|
| | | | | | | 1999 Chi-Chi $M_w$ 7.6 | 2008 Wenchuan $M_w$ 7.9 | 2011 Tohoku $M_w$ 9.0 |
| Water radon concentration | FD-125 | Beijing Nuclear Instrument Factory | 1986 | 1 day | 10% | Rise, ~ 14%, ~ 1 year | Rise, ~ 22%, ~ 4.5 month | Rise, ~ 12%, ~ 8 month |
| Water pressure | Manometer | Shanghai Automation Instrument Co., Ltd. | 1986 | 1 day | 0.4% | Rise, ~ 6.2%, ~ 6 month | Rise, ~ 7.0%, ~ 4.5 month | Rise, ~ 3.8%, ~ 8 month |
| Discharge | Flowmeter | Dongying Bangrui Instrument Co., Ltd. | 1986 | 1 day | 0.5% | No change | Rise, ~ 1.2%, ~ 4.5 month | No change |
| Water level | RGP-1 | Institute of Earthquake Science, CEA | 1996 | 1 h | 0.0001 m | Rise, ~ 0.18 m, ~ 6 month | | |
| | LN-3A | Institute of Earthquake Science, CEA | 2000 | 1 min | 0.2% of the full-scale | | Rise (with oscillations), 0.19 m (1.02 m), ~ 4.5 month | Oscillations, 0.91 m, < 1 day |
| Water temperature | Mercury thermometer | | 1986 | 1 day | 0.1 °C | No change | No change | No change |
| | SZW-1A | Institute of Crustal Dynamics, CEA | 1996 | 1 h before 2000,1 min after 2000 | < 0.05 °C | Drop, 0.034 °C | Drop, 0.044 °C, ~ 4.5 month | Drop, 0.014 °C |

The RGP-1 was not used after January 2000. The amplitudes of the water radon concentration, water pressure, and discharge changes were roughly estimated because their background variations are significant. The amplitudes of the water temperature changes recorded by the SZW-1A are lower than the observation accuracy but changes are resolvable because the resolution of the SZW-1A is 0.0001 °C and amplitudes of the temperature changes and the observation accuracy have the similar magnitude. The recovery time of the water temperature changes recorded by the SZW-1A after the Chi–Chi and the Tohoku earthquakes is not known, because the steps due to the instrument malfunctions after the two earthquakes (on 3 October 1999 and 1 April 2011 for the Chi–Chi and the Tohoku earthquakes, respectively) made it difficult to distinguish them

$$C = \frac{K(N - N_0)}{V \cdot e^{-\lambda t}} \qquad (1)$$

where $K$ is the calibration value of the scintillation chamber, $N_0$ is the background pulse frequency of the scintillation chamber, $V$ is the volume of the sampling water, and $e^{-\lambda t}$ is the decay constant of radon. The $K$ is calibrated once a year by a solid radium source whose radon content is known. Water radon concentrations of two samples collected at the same time were compared between 2000 and 2007, and the results are consistent.

## 3. Hydrological Changes

The hydrological data at the Lujiang well from January 1987 to December 2016 were collected for studying the earthquake-induced hydrological changes. Regional earthquakes did not cause detectable hydrological changes, as the seismic activity near the Lujiang well is relatively low since 1980, and the biggest local event is the 2005 Jiujiang $M_w$ 5.2 earthquake, which is 220 km away from the Lujiang well. Hydrological changes were detected following 74 distant earthquakes (some of them are shown in Fig. 1). The water level responded to all of these distant earthquakes, rises for the 1999 Chi–Chi $M_w$ 7.6, the 2004 Sumatra $M_w$ 9.0 and the 2008 Wenchuan $M_w$ 7.9 earthquakes, and oscillates for other earthquakes. The water radon concentration, water pressure, and discharge rose after several of these earthquakes, whereas the water temperature dropped. Sustained changes were recorded after the Chi–Chi, the 2007 Sumatra $M_w$ 8.5, the Wenchuan and the 2011 Tohoku $M_w$ 9.0 earthquakes.

The seismic energy density was used to relate and compare the hydrological changes of different parameters at the Lujiang well. The seismic energy density ($e$) is calculated by the formula (Wang and Manga 2010b):

$$\log(r) = 0.48M - 0.33 \log e(r) \qquad (2)$$

where $r$ is the epicenter distance and $M$ is the earthquake magnitude. The calculation shows that the threshold seismic energy densities required to initiate the changes in water level and water temperature are

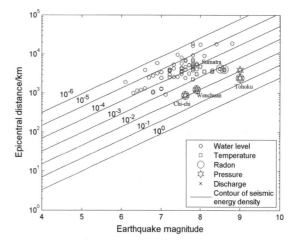

Figure 3
Graph of the magnitude–distance and the hydrological responses at the Lujiang well. The graph is based on the model of Wang and Manga (2010b). The hydrological responses whose durations exceeded 2 days are featured with place names where relevant earthquakes happened. Two days is the least resolvable time for the observation parameters whose sampling intervals are 1 day. Some of the earthquakes in this figure are shown in Fig. 1. The moment magnitude ($M_w$) is used. The unit of the seismic energy density is $J/m^3$

$10^{-6}$ and $10^{-4}$ $J/m^3$, respectively. The threshold seismic energy density for the water radon concentration, water pressure, and discharge changes is $10^{-3}$ $J/m^3$ (Fig. 3).

In this study, we focused on the hydrological changes induced by three distant earthquakes, i.e., the Chi–Chi, the Wenchuan and the Tohoku earthquakes, which are 879, 1233 and 2473 km away from the Lujiang well, respectively. The three earthquakes caused sustained changes in several observation parameters. The hydrological changes induced by the three earthquakes are summarized in Table 1 and are graphed in Figs. 4, 5, 6. As the figures show, these hydrological changes are synchronous and have similar amplitudes. The responses following the Chi–Chi earthquake were rapid, whereas those following the Wenchuan and Tohoku earthquakes were much slower. These changes lasted for several months, except for the water level oscillations after the Tohoku earthquake, which lasted only several hours. Those water level oscillations are different from the water pressure rise following the Tohoku earthquake, though both the water level and pressure were

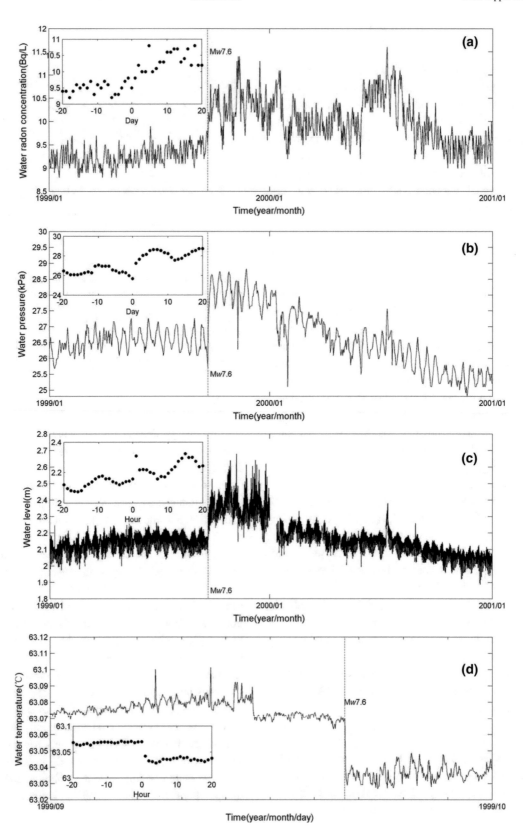

◀Figure 4
Hydrological changes at the Lujiang well following the Chi–Chi earthquake. The dotted red lines denote the start time of the earthquake. The insets are magnified views of the changes before and after the earthquake time (time 0)

measured via pressure-based recorders; we have no reasonable explanation.

The location where these hydrological changes took place is learned by a simple estimation. The discharge ($\sim$ 7.35 L/s) and the diameter ($\sim$ 14 cm) of the Lujiang well imply that the flow rate is $\sim$ 0.5 m/s. The water temperature changes lagged behind that of the water level for $\sim$ 5 min after the Wenchuan and the Tohoku earthquakes. The flow rate and the time lags imply that the distance between the place where the hydrological changes happened and the place where the temperature was measured (34.5 m) is $\sim$ 150 m. The estimation shows that these hydrological changes happened adjacent to the wellbore.

## 4. Tidal Analysis

The analysis of the responses of well water level to Earth tides is a probe to monitor aquifer permeability, both for the horizontal flow (Hsieh et al. 1987; Elkhoury et al. 2006) and for the vertical flow (Allègre et al. 2016; Xue et al. 2016). The $M_2$ wave is commonly used because of its greater amplitude than other waves and because it is less contaminated by the barometric pressure or diurnal temperature fluctuations (Lai et al. 2014).

We used the hourly water level data of the Lujiang well for tidal analysis. With a 7-day sliding window, we divided the data into different segments with each segment of 30 days. The lengths of the sliding window and the segment are based on Wang et al. (2016), which discusses the effect of different lengths. We extracted the amplitude and phase responses of different segments at the frequency of the $M_2$ wave by the VAV program (Venedikov et al. 2003).

The results are displayed in Fig. 7. The phase shift of water level to Earth tides is positive, and a positive

phase shift is an indication that there was significant vertical flow (Roeloffs 1996). The vertical flow is further supported by the responses of water temperature to Earth tides and by the changes of water temperature after earthquakes. The water temperature has the synchronous oscillations of the water level, and the amplitude spectrum indicates that these changes are correlated with Earth tides (Fig. 8). A similar phenomenon has been reported and is considered to be originated from the groundwater flow (Furuya and Shimamura 1988; Mogi et al. 1989; Yu et al. 1997). In the case of the Lujiang well, the observed water temperature ($\sim$ 63 °C) and the temperature of the hottest aquifer (62.5 °C) are approximately the same (Fig. 2), indicating that the temperature gradient inside the wellbore is too small to account for the water temperature oscillations. Therefore, the water temperature response to Earth tides originates from vertical flow in its aquifers. The oscillations of the water temperature and the water level are synchronous under the forces of Earth tides (Fig. 8); however, the water temperature dropped whereas the water level rose after earthquakes (Figs. 4, 5, 6). The transformation suggests that significant vertical flow happened after earthquakes.

The permeability inferred through the tidal response of water level was calculated using a vertical flow model. The method is based on Allègre et al. (2016) and Xue et al. (2016), which calculated permeability when the phase response is between $-1°$ and $45°$. The amplitude response ($A$) and the phase response ($\eta$) can be expressed as:

$$A = \frac{1}{S_s}\left[1 - 2\exp\left(-\frac{z}{\delta}\right)\cos\frac{z}{\delta} + \exp\left(-\frac{2z}{\delta}\right)\right]^{1/2} \tag{3}$$

$$\eta = \tan^{-1}\left\{\frac{\exp\left(-\frac{z}{\delta}\right)\sin\frac{z}{\delta}}{1 - \exp\left(-\frac{z}{\delta}\right)\cos\frac{z}{\delta}}\right\} \tag{4}$$

where $S_s$ is the specific storage, $z$ is the depth from the surface, $\delta = \sqrt{\frac{2D}{w}}$, $D$ is the hydraulic diffusivity which equals the transmissivity $T$ divided by the storativity $S$; $w$ is the angular frequency of the oscillation. $T$ and $S$ can be estimated by fitting Eqs. (3) and (4) to the measured amplitude and phase

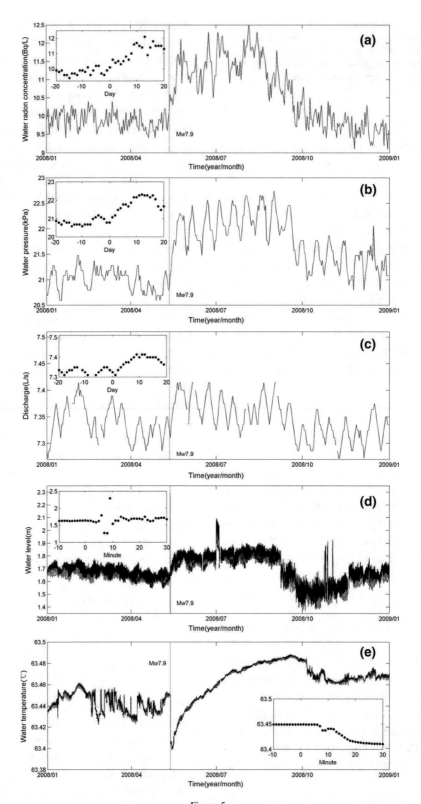

Figure 5
Hydrological changes at the Lujiang well following the Wenchuan earthquake

Hydrological Changes Induced by Distant Earthquakes at the Lujiang Well in Anhui, China

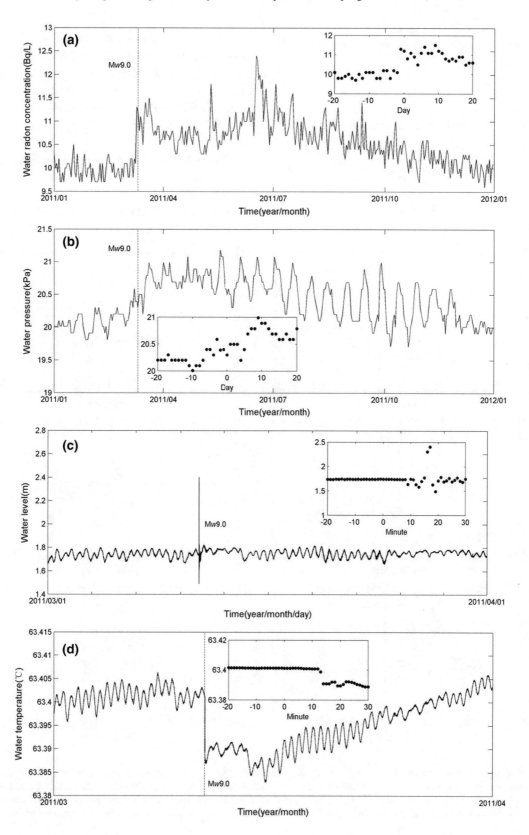

◀Figure 6
Hydrological changes at the Lujiang well following the Tohoku
earthquake

where $\mu$ is the fluid dynamic viscosity, $\rho$ is the density of the fluid, g is the gravitational acceleration, $b$ is the length of the open interval of the well. For the Lujiang well, we used $z = 42$ m, $w = 1.4053 \times 10^{-4}$ rad/s, $\mu = 0.447 \times 10^{-3}$ Pa s, $\rho = 0.98163 \times 10^{3}$ kg/m$^3$, $g = 9.8$ m/s$^2$ and $b = 285$ m.

responses. $T$ and $S$ can be converted to permeability ($k$) by:

$$k = \frac{\mu}{\rho g b} T \qquad (5)$$

$$S_s = S/b \qquad (6)$$

The calculated permeability at the Lujiang well before and after the Chi–Chi, the Wenchuan and the Tohoku earthquakes are shown in Fig. 9. It indicates that the permeability showed insignificant change

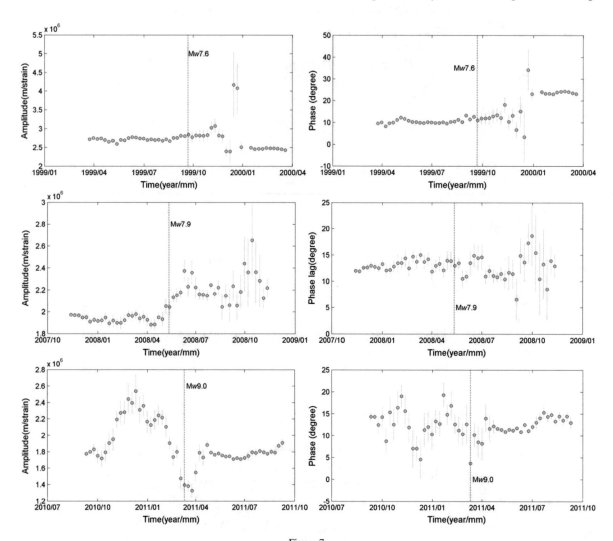

Figure 7
Amplitude response (left) and phase response (right) of the $M_2$ wave at the Lujiang well. The amplitude response is the amplitude of the water level relative to solid Earth tides. The phase response is the phase shift of the water level relative to solid Earth tides, and a positive value means the water level precedes Earth tides. The figures at the top, the middle and the bottom represent the responses before and after the Chi–Chi earthquake, the Wenchuan earthquake, and the Tohoku earthquake, respectively. The bars denote the mean square errors

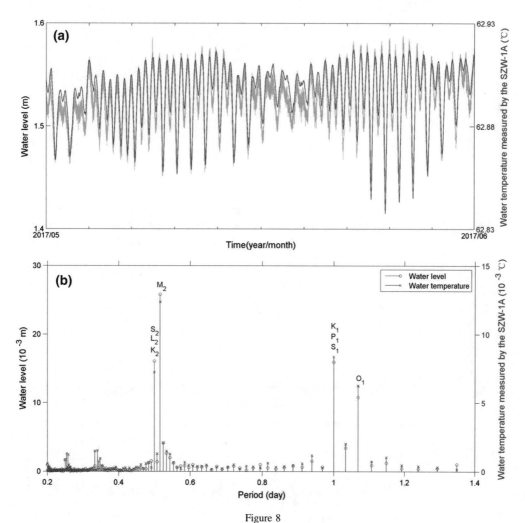

Figure 8
Raw water temperature data and water level data at the Lujiang well (**a**), as well as the amplitude spectrum of the data (**b**). The texts in (**b**) denote the names of tidal waves

after the three earthquakes. Thus, the permeability enhancement cannot explain the hydrological changes.

## 5. Discussion

The mechanism responsible for the hydrological changes was further studied by surveying the features of the wells which are adjacent to the Lujiang well. Basic information on the adjacent wells is provided in Table 2. An adjacent well (the #3 well) shares the same aquifer with the Lujiang well because the two wells have hydraulic connectivity. The lithologic logs, the water temperature data, and the chemical analyses further support this (see Table S1 and Figs. S1–S5 in the Supplementary material). The #3 well has similar water temperature and similar water radon concentration with those of the Lujiang well. The water radon concentration of the Lujiang well increased as the #3 well was closed (Fig. 10), suggesting that the water radon concentration of the Lujiang well increased following the increase of recharge to the Lujiang well. Another two adjacent wells (the #7 and #11 wells) that have different aquifers with the Lujiang well, either shallower or

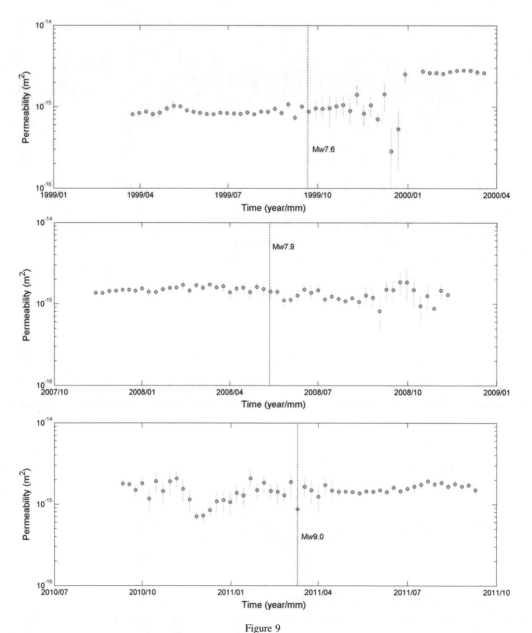

Figure 9
Permeability variations at the Lujiang well before and after the Chi–Chi, the Wenchuan, and the Tohoku earthquakes. The bars denote the mean square errors

deeper than the Lujiang well, have higher water radon concentrations and lower water temperatures than those of the Lujiang well.

Therefore, the increase in the vertical recharge seems to be a considerable mechanism responsible for the hydrological changes. Though both the recharge from the shallower and deeper aquifers might lead to the rise in the water radon concentration and the drop in the water temperature at the Lujiang well, we were more inclined to the deeper recharge on the basis that the Lujiang well was affected by the massive pumping of a deeper well (the #11 well: Fig. 11).

Table 2

*Basic information of the Lujiang well and its adjacent wells*

| Well name | Well type | Depth (m) | Aquifer lithology | Hydrochemical type of groundwater | Distance from the Lujiang well (m) | Water radon concentration (Bq/L) | Water temperature (°C) |
|---|---|---|---|---|---|---|---|
| Lujiang well | Artesian well | 327.05 | Tuff, diorite-porphyrite | Sodium sulfate | | ~ 10 | ~ 63 |
| #3 well | Artesian well | 145.85 | Tuff, diorite-porphyrite | Sodium sulfate | ~ 150 | ~ 9 | ~ 63 |
| #7 well | Pumping well | 253.60 | Tuff | Sodium sulfate | ~ 200 | ~ 40 | ~ 37 |
| #11 well | Pumping well | 932.31 | Tuff | Sodium bicarbonate | ~ 650 | ~ 34 | ~ 42 |

The Lujiang well is also known as the #1 well. The water radon concentration of the #3 well and the #11 well is from Liu et al. (2010), and the water concentration of the #7 well is from the Lujiang station. The water radon concentration was measured by the same instrument and the staff as those of the Lujiang well. Other information is from Li et al. (2010). The water temperatures of the Lujiang well and the #3 well are the hottest in the Tangchi Town

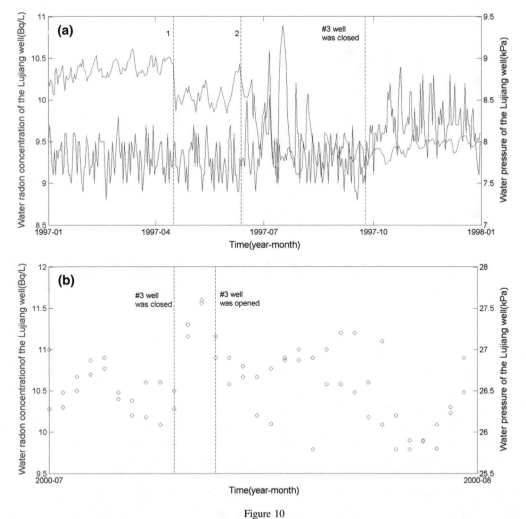

Figure 10

**a, b** Water radon concentration of the Lujiang well increased after the #3 well was closed. The #3 well was drilled on 16 April 1997 (the dotted line marked with number 1) and its initial discharge is 80 m³ per day (m³/day), the discharge was increased to more than 250 m³/day between 21 June 1997 (number 2) and late September 1997. These messages are from Li et al. (2010)

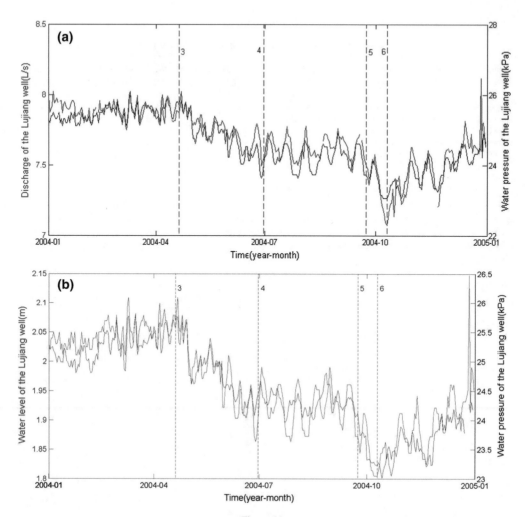

Figure 11
**a**, **b** Lujiang well was affected by the massive pumping of the #11 well. The pumping rate of the #11 well was increased from fewer than 250 m³/day to more than 400 m³/day between late April 2004 (number 3) and 30 June 2004 (number 4). The pumping rate of the #11 well was increased to more than 400 m³/day between 23 September 2004 (number 5) and early October 2004 (number 6). These messages are from Li et al. (2010)

The hot water from the Lujiang well is an important hydrothermal resource in the Lujiang County; however, the amount, temperature, and chemical composition of the water were significantly changed after several distant earthquakes. The increase in vertical recharge may cause the changes. Thus, the vertical recharge enhancement induced by the far-field earthquakes should be considered in other cases of studies.

## 6. Conclusions

We present the sustained changes of different parameters induced by three distant earthquakes at a hot artesian well. These hydrological changes are synchronous and have similar amplitudes. We use the tidal response of water level to show that the earthquake-enhanced permeability cannot explain the changes. Based on the hydrological observations at the Lujiang well and the features of its adjacent wells, we attribute the changes to the increase in the vertical recharge.

## Acknowledgements

We thank the Guest Editor, Michael Manga, and the two anonymous reviewers for their valuable comments that helped much for improving this paper. The data in this study were from the China Earthquake Networks Center. This study is supported by the National Natural Science Foundation of China (U1602233, 41572238, 41602274) and the Spark Program of Earthquake Science of China (XH17045). We thank Chuanqin Liu for providing the information of the #11 well. The earthquake information in this paper was obtained from the website: http://www.globalcmt.org/CMTsearch.html.

## REFERENCES

Allègre, V., Brodsky, E. E., Xue, L., Nale, S. M., Parker, B. L., & Cherry, J. A. (2016). Using earth-tide induced water pressure changes to measure in situ permeability: A comparison with long-term pumping tests. *Water Resources Research, 52,* 3113–3126.

Anhui Earthquake Administration. (2004). *Seismic monitoring record of Anhui Province* (pp. 81–93). Hefei: Anhui University Press. (**in Chinese**).

Bo, L., Sun, X.W., & Sun, X. (2007). Geothermal in Tangchi Town, Lujiang County, Anhui Province and its exploitation and utilization. *Proceedings in earth science in the East China region in 2007.* Hefei University of Technology Press, Hefei, pp: 178-183. (**in Chinese**).

Brodsky, E. E., Roeloffs, E. A., Woodcock, D., Gall, I., & Manga, M. (2003). A mechanism for sustained groundwater pressure changes induced by distant earthquakes. *Journal of Geophysical Research, 108,* 2390.

Claesson, L., Skelton, A., Graham, C., Dietl, C., Mörth, M., Torssander, P., et al. (2004). Hydrogeochemical changes before and after a major earthquake. *Geology, 32,* 641–644.

Cooper, H. H., Bredehoeft, J. D., Papadopulos, I. S., & Bennett, R. R. (1965). The response of well-aquifer systems to seismic waves. *Journal of Geophysical Research, 70,* 3915–3926.

Elkhoury, J. E., Brodsky, E. E., & Agnew, D. C. (2006). Seismic waves increase permeability. *Nature, 441,* 1135–1138.

Furuya, I., & Shimamura, H. (1988). Groundwater microtemperature and strain. *Geophysical Journal International, 94*(2), 345–353.

Geballe, Z. M., Wang, C.-Y., & Manga, M. (2011). A permeability-change model for water-level changes triggered by teleseismic waves. *Geofluids, 11,* 302–308.

Hsieh, P. A., Bredehoeft, J. D., & Farr, J. M. (1987). Determination of aquifer transmissivity from Earth tide analysis. *Water Resources Research, 23,* 1824–1832.

Huang, F. Q., Jian, C. L., Tang, Y., Xu, G. M., Deng, Z. H., & Chi, G. C. (2004). Response changes of some wells in the mainland subsurface fluid monitoring network of China, due to the September 21, 1999, $M_s$ 7. 6 Chi-Chi earthquake. *Tectonophysics, 390,* 217–234.

King, C. Y., Azuma, S., Igarashi, G., Ohno, M., Saito, H., & Wakita, H. (1999). Earthquake-related water-level changes at 16 closely clustered wells in Tono, central Japan. *Journal of Geophysical Research: Solid Earth, 104,* 13073–13082.

Lai, G. J., Ge, H. K., Xue, L., Brodsky, E. E., Huang, F. Q., & Wang, W. L. (2014). Tidal response variation and recovery following the Wenchuan earthquake from water level data of multiple wells in the near field. *Tectonophysics, 619–620,* 115–122.

Lai, G. J., Jiang, C. S., Han, L. B., Sheng, S. Z., & Ma, Y. C. (2016). Co-seismic water level changes in response to multiple large earthquakes at the LGH well in Sichuan, China. *Tectonophysics, 679,* 211–217.

Li, Z. P., Zhu, F. B., Wang, S. X., Shu, K. M., & Chen, H. B. (2010). Influence of exploitation of hot water well on the fluids dynamics in Lujiang seismic station. *Seismological and geomagnetic observation and research, 31,* 91–96. (**in Chinese**).

Liao, X., Wang, C. Y., & Liu, C. P. (2015). Disruption of groundwater systems by earthquakes. *Geophysical Research Letters, 42,* 9758–9763.

Liu, L. B., Roeloffs, E. A., & Zheng, X. Y. (1989). Seismically induced water level fluctuations in the Wali well, Beijing, China. *Journal of Geophysical Research: Solid Earth, 94,* 9453–9462.

Liu, C. Q., Wang, S. X., Zhu, F. B., Shu, K. M., Zhou, Z. G., Pei, H. Y., et al. (2010). Analysis on the reason for the high-value anomaly of water radon at No.1 well in Tangchi. *Seismological and geomagnetic observation and research, 33,* 154–158. (**in Chinese**).

Ma, Y. C., & Huang, F. Q. (2017). Coseismic water level changes induced by two distant earthquakes in multiple wells of the Chinese mainland. *Tectonophysics, 694,* 57–68.

Manga, M., Beresnev, I., Brodsky, E. E., Elkhoury, J. E., Elsworth, D., Ingebritsen, S. E., et al. (2012). Changes in permeability caused by transient stresses: Field observations, experiments, and mechanisms. *Reviews of Geophysics, 50*(2), RG2004.

Mogi, K., Mochizuki, H., & Kurokawa, Y. (1989). Temperature changes in an artesian spring at Usami in the Izu Peninsula (Japan) and their relation to earthquakes. *Tectonophysics, 159,* 95–108.

Ren, H. W., Liu, Y. W., & Yang, D. Y. (2012). A preliminary study of post-seismic effects of radon following the $M_s$ 8.0 Wenchuan earthquake. *Radiation Measurements, 47,* 82–88.

Roeloffs, E. A. (1996). Poroelastic techniques in the study of earthquake-related hydrologic phenomena. *Advances in Geophysics, 37,* 135–195.

Roeloffs, E. A. (1998). Persistent water level changes in a well near Parkfield, California, due to local and distant earthquakes. *Journal of Geophysical Research, 103,* 868–889.

Rojstaczer, S., Wolf, S., & Michel, R. (1995). Permeability enhancement in the shallow crust as a cause of earthquake-induced hydrological changes. *Nature, 373,* 237–239.

Shi, Y. L., Cao, J. L., Ma, L., & Yin, B. J. (2007). Tele-seismic coseismic well temperature changes and their interpretation. *Acta Seismologica Sinica, 20,* 280–289.

Shi, Z. M., & Wang, G. C. (2014). Hydrological response to multiple large distant earthquakes in the Mile well, China. *Journal of Geophysical Research: Earth Surface, 119,* 2448–2459.

Shi, Z. M., & Wang, G. C. (2016). Aquifers switched from confined to semiconfined by earthquakes. *Geophysical Research Letters, 43,* 11166–11172.

Shi, Z. M., Wang, G. C., Manga, M., & Wang, C.-Y. (2015). Mechanism of co-seismic water level change following four great earthquakes—insights from co-seismic responses throughout the Chinese mainland. *Earth and Planetary Science Letters, 430,* 66–74.

Skelton, A., Andrén, M., Kristmannsdóttir, H., Stockmann, G., Mörth, C. M., Sveinbjörnsdóttir, Á., et al. (2014). Changes in groundwater chemistry before two consecutive earthquakes in Iceland. *Nature Geoscience, 7,* 752–756.

Sun, X. L., Wang, G. C., & Yang, X. H. (2015). Coseismic response of water level in Changping well, China, to the $M_w$ 9.0 Tohoku earthquake. *Journal of Hydrology, 531,* 1028–1039.

Venedikov, A. P., Arnoso, J., & Vieira, R. (2003). VAV: A program for tidal data processing. *Computers and Geosciences, 29,* 487–502.

Wakita, H., Igarashi, G., Nakamura, Y., Sano, Y., & Notsu, K. (1989). Coseismic radon changes in groundwater. *Geophysical Research Letters, 16,* 417–420.

Wang, C.-Y., Chia, Y.-P., Wang, P.-L., & Dreger, D. (2009). Role of S waves and Love waves in coseismic permeability enhancement. *Geophysical Research Letters, 36,* L09404.

Wang, C.-Y., Liao, X., Wang, L.-P., Wang, C.-H., & Manga, M. (2016). Large earthquakes create vertical permeability by breaching aquitards. *Water Resources Research, 52*(8), 5923–5937.

Wang, C.-Y., & Manga, M. (2010a). *Earthquakes and water* (pp. 67–99). Berlin: Springer.

Wang, C.-Y., & Manga, M. (2010b). Hydrologic responses to earthquakes and a general metric. *Geofluids, 10*(1–2), 206–216.

Wang, C.-Y., Manga, M., Wang, C.-H., & Chen, C.-H. (2012). Transient change in groundwater temperature after earthquakes. *Geology, 40,* 119–122.

Wang, C.-Y., Wang, C.-H., & Manga, M. (2004). Coseismic release of water from mountains: Evidence from the 1999 ($M_w$ = 7.5) Chi–Chi, Taiwan, earthquake. *Geology, 32,* 769–772.

Wang, C.-Y., Wang, L.-P., Manga, M., Wang, C.-H., & Chen, C.-H. (2013). Basin-scale transport of heat and fluid induced by earthquakes. *Geophysical Research Letters, 40,* 3893–3897.

Weingarten, M., & Ge, S. M. (2014). Insights into water level response to seismic waves: A 24 year high-fidelity record of global seismicity at Devils Hole. *Geophysical Research Letters, 41,* 74–80.

Xue, L., Brodsky, E. E., Erskine, J., Fulton, P. M., & Carter, R. (2016). A permeability and compliance contrast measured hydrogeologically on the San Andreas Fault. *Geochemistry, Geophysics, Geosystems, 17,* 858–871.

Xue, L., Li, H. B., Brodsky, E. E., Xu, Z. Q., Kano, Y., Wang, H., et al. (2013). Continuous permeability measurements record healing inside the Wenchuan earthquake fault zone. *Science, 340,* 1555–1559.

Yan, R., Woith, H., & Wang, R. J. (2014). Groundwater level changes induced by the 2011 Tohoku earthquake in China mainland. *Geophysical Journal International, 199,* 533–548.

Yu, J. Z., Che, Y. T., & Liu, W. Z. (1997). Preliminary study on the hydrodynamic mechanism of micro behavior of water temperature in well. *Earthquake, 17,* 389–396. **(in Chinese)**.

Zhang, C. M. (1991). Design of diversion water sampling system of an artesian well at the Lujiang synthetic hydrochemical station. *Earthquake, 1,* 59–65. **(in Chinese)**.

Zhang, P. Z., Deng, Q. D., Zhang, G. M., Ma, J., Gan, W. J., Min, W., et al. (2003). Active tectonic blocks and strong earthquakes in the continent of China. *Science in China, Series D: Earth Sciences, 46,* 13–24.

(Received January 16, 2017, revised October 16, 2017, accepted October 25, 2017)

Pure Appl. Geophys.
© 2017 Springer International Publishing AG
DOI 10.1007/s00024-017-1643-6

# Quantitative Assessment of the Mechanisms of Earthquake-Induced Groundwater-Level Change in the MP Well, Three Gorges Area

SHOUCHUAN ZHANG,[1] ZHEMING SHI,[1,2] GUANGCAI WANG,[1,2] and ZUOCHEN ZHANG[3]

*Abstract*—Earthquake-induced groundwater-level changes have been widely studied, though the mechanisms causing coseismic responses are still debated. In this study, we employ several models to fit the coseismic groundwater-level changes caused by the 2008 Wenchuan earthquake in the MP well, located in the Three Gorges Dam. The fits for all models are about the same. By comparing the model results with the results from tidal response and baseflow recession analyses, we conclude that a transient permeability model can best describe the coseismic groundwater-level changes in the MP well. The discharge from the Changmutuo fault zone estimated from the one-dimensional groundwater flow model during the 20 days following the earthquake is about $310 \pm 90$ m$^3$.

**Key words:** Coseismic, water level, permeability, mechanism, model.

## 1. Introduction

Groundwater disturbances following earthquakes have been widely noted and discussed in many studies (Muir-Wood and King 1993; Roeloffs 1998; Shi and Wang 2015; Shi et al. 2014; Sun et al. 2015; Wang et al. 2004c). There are several reasons to investigate earthquake-induced groundwater-level changes. First, these changes may reflect the disruption of aquifer systems (Liao et al. 2015; Shi and Wang 2016; Wang et al. 2016) and may affect water supplies (Gorokhovich 2005; Gorokhovich and

Fleeger 2007; Rojstaczer et al. 1995). Second, the groundwater-level changes following earthquakes reflect the redistribution of pore pressure which may influence aftershocks (Bosl and Nur 2002). Third, the changes may also put some underground waste repositories at risk (Carrigan et al. 1991).

There are currently several mechanisms proposed to explain the coseismic groundwater-level disturbances: (1) coseismic elastic strain model (Ge and Stover, 2000); (2) undrained consolidation of sediments (Wang et al. 2001); (3) permeability changes of the aquifers (Brodsky et al. 2003; Elkhoury et al. 2006; Geballe et al. 2011; Manga et al. 2012; Shi et al. 2015a). Among these mechanisms, permeability changes have been the most widely favored mechanism (Shi et al. 2015a). Analytical models have been used to quantitatively evaluate the coseismic water-level changes caused by permeability changes (Geballe et al. 2011; Roeloffs et al. 2003; Rojstaczer et al. 1995; Tokunaga 1999; Wang et al. 2004b). Continuous field observation of permeability changes before and after earthquakes have also been documented by analyzing the tided response of wells (Elkhoury et al. 2006; Lai et al. 2016; Shi and Wang 2014; Xue et al. 2013; Yan et al. 2016).

However, there are no studies that use both groundwater-level changes and tidal responses to confirm the permeability mechanism. In analytical studies, aquifer parameters before and after earthquakes are inferred from the models by fitting the water-level data (Geballe et al. 2011; Manga and Rowland 2009; Tokunaga 1999; Wang et al. 2012). However, a good fit to the field data does not prove the validity of the model; it merely demonstrates the consistency of the model with the field data (Wang et al. 2004c). For the tidal analysis, although the

---
[1] State Key Laboratory of Biogeology and Environmental Geology, MOE Key Laboratory of Groundwater Circulation and Environmental Evolution, China University of Geosciences, Beijing 100083, China. E-mail: szm@cugb.edu.cn
[2] School of Water Resources and Environment, China University of Geosciences, Beijing 100083, China.
[3] Geological Environmental Monitoring Institute of China, Beijing 100081, China.

permeability evolution before and after earthquakes could be calculated (Elkhoury et al. 2006; Shi and Wang 2015; Yan et al. 2016), no independent field data had been used to evaluate these results. Thus, the mechanisms of permeability and coseismic groundwater-level changes are still debated. In this study, we combine a model for the groundwater-level change with the tidal response to constrain the mechanism of coseismic changes. We choose a groundwater monitoring well that records clear tidal signals and coseismic responses to shed light on the interaction between earthquake and a well–aquifer system.

## 2. The Observational Wells and Their Response to the Earthquake

### 2.1. Geological Setting of the Well–Aquifer System

The groundwater well, MP, is part of the Three Gorges groundwater observational network. It was established to monitor possible reservoir-induced seismicity in the vicinity of the Three Gorges Dam. As shown in Fig. 1, the MP well is located 50 m west of the central section of the Changmutuo Fault and connected with the fault zone. It is constructed in a quartz diorite aquifer with depth of 200.5 m. Data

have been collected since July 2001. Figure 2 shows the screen depth and surrounding rocks. The well has a diameter of 114 mm, with a screen interval between 85 and 130 m. The specific storage ($S_s$) is $1.7 \times 10^{-5}$ $1$ m$^{-1}$ with hydraulic conductivity ($K$) of 0.01 m day$^{-1}$ measured by a pump test (Che et al. 2002). The measured water-level changes correspond to the height of the water column between the sensor and water surface. Water levels are recorded by a DSW-01 digital instrument developed by the Institute of Seismology, China Earthquake Administration (CEA). The measurement range is from 0 to 10 m with a resolution of 1 mm and sampling rate of 1 min (Shi et al. 2013).

### 2.2. Properties and Coseismic Water-Level Response of the MP Well

According to the water sample collected in 2002, the chemical type of groundwater is HCO$_3$–Ca–Na;

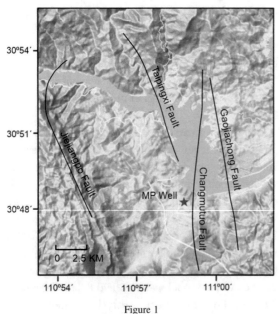

Figure 1
Location of the MP well

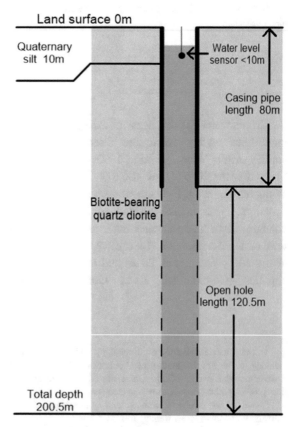

Figure 2
Aquifer lithology and specification of the well–aquifer system

$\delta D = -59.96‰$, $\delta^{18}O = -8.49‰$, $\delta^{13}C = -12.51‰$ and $^3H = 0$ TU (Che et al. 2002). These chemical and isotopic characteristic indicate that the groundwater is recharged from precipitation and has a long residence time (>50 year).

On May 12, 2008, the $M_s$ 8.0 Wenchuan earthquake caused continental-scale coseismic groundwater-level changes (Shi et al. 2015b). For the MP well, groundwater level showed a sustained drop with a maximum of 0.178 m that lasted more than a month before returning to the level before the earthquake. In a previous study, Shi et al. (2013) used the tidal response to show that the transmissivity of the MP well–aquifer system increased by 70% following the Wenchuan earthquake and then recovered. However, no other support was provided for their conclusion. To further understand the mechanism of the coseismic water-level changes, we employ several models to evaluate the coseismic groundwater changes mechanism.

## 3. Models to Explain the Coseismic Groundwater-Level Changes

Several models have been successfully used to explain earthquake-induced water-level or discharge changes (Manga et al. 2003; Manga and Rowland 2009; Roeloffs 1998; Rojstaczer et al. 1995; Tokunaga 1999; Wang et al. 2004b). The MP well is located in the intermediate field (with epicenter distance more than one ruptured fault length), so the coseismic elastic strain model is not relevant. Since the MP well is located in crystalline rocks, consolidation/liquefaction is also not possible. As the MP well is constructed in a fault zone and close to the reservoir, both permeability change in the fault zone and increased recharge from the reservoir are possible. However, there is no hydraulic connection between the MP well and the reservoir according to CFCs studies (Zhang et al. 2014). Together with the previous tidal response study, we assume that changes in the hydraulic properties of the fault zone are a more plausible mechanism. Here, we follow the fault zone models proposed by Manga and Rowland (2009) to explain our data. These models are (Fig. 3): (a) enhanced permeability model; (b) transient permeability model; (c) increased head

model; (4) one-dimensional groundwater flow model. The fault zone in these models is treated as a homogenous one-dimensional aquifer. Although idealized, these models fit the observed data well (Manga and Rowland 2009).

### 3.1. Enhanced Permeability Model

The enhanced permeability model assumes that the hydraulic conductivity in the fault zone increases following the earthquake. The coseismic changes of hydraulic conductivity cause coseismic changes in discharge and also change the water level (Fig. 3a). The evolution of head in the fracture system can be approximated by the groundwater flow equation with an additional term that accounts for recharge to the fracture zone (Manga and Rowland 2009):

$$S_s \frac{\partial h}{\partial t} = K_f \frac{\partial^2 h}{\partial x^2} + \frac{K_f}{wD}(h_0 - h), \quad (1)$$

with boundary condition

$$h = h_0, \text{ at } z = D. \quad (2)$$

Here, $S_s$ is the storage property of the fault zone, $K_f$ the hydraulic conductivity of the fault zone, $w$ the width of the fault zone and $D$ the distance from the fault to the boundary where the head is fixed to $h_0$. Fitting the results of the equation to the measured water-level changes, the hydraulic conductivity in the fault zone can be obtained. In this model, we assume that specific storage, $S_s$, does not change, though both hydraulic conductivity and storage properties can be influenced by earthquakes.

This model is characterized by four parameters:

$$S_s; h_0; K_f; \alpha = \frac{K_f}{wD},$$

where $S_s$ is the specific storage of the fault zone according to the pump test and $h_0$ is the water level measured before the earthquake. Both $K_f$, the hydraulic conductivity of the fracture zone, and $\alpha$, a scaling parameter, are fit in this model.

### 3.2. Transient Permeability Model

As suggested by previous studies, the permeability changes following an earthquake may not be

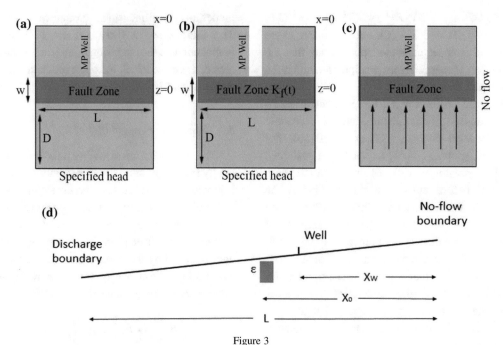

Figure 3
Schematic illustration of conceptual models: **a** enhanced permeability model; **b** transient permeability model; **c** increased head model; **d** Roeloffs's one-dimensional groundwater flow model

permanent and will recover gradually through poroelastic processes or geochemical/biogeochemical processes (Manga et al. 2012). We assume that the hydraulic conductivity changes from the pre-earthquake ($K_{fi}$) to the post-earthquake ($K_{ff}$) (Fig. 3b), exponentially in time with decay constant $\lambda$. The hydraulic conductivity can be expressed as:

$$K_f(t) = K_{fi} + (K_{ff} - K_{fi})e^{-\lambda t}, \qquad (3)$$

where $K_f$ is the hydraulic conductivity of the fracture zone and the subscripts '$i$' and '$f$' indicate values before (initial) and after (final) the earthquake. With the boundary condition $h = h_0$ at $z = D$, the groundwater flow is

$$S_s\frac{\partial h}{\partial t} = \left[K_{fi} + (K_{ff} - K_{fi})e^{-\lambda t}\right]\frac{\partial^2 h}{\partial x^2} + \frac{K_f}{wD}(h_0 - h).$$
$$(4)$$

This model is characterized by six parameters:

$$S_s;\ h_0;\ K_1 = K_{fi};\ K_2 = K_{ff};\ \beta = \frac{K_f}{wD};\ \lambda.$$

Again, $S_s$ can be obtained from the pump test, and $h_0$ is the water level before the earthquake. $K_1$ and $K_2$ are the hydraulic conductivity of the fault before and after the earthquake. $\beta$ is a scaling parameter. $K_1$, $K_2$, $\beta$ and $\lambda$ are the parameters fit in this model.

### 3.3. Increased Head Model

If $K_f$ of the fault zone remains unchanged, but permeability in the surrounding region changes by breaching of barriers, recharge/discharge to the well will change (Wang et al. 2004a, b) (Fig. 3c). In this case, there is no change in the hydraulic conductivity in the fault zone after the earthquake, and the cross-sectional area $A$ of the fault does not change. It only causes the increase/decrease of water supply around the fault:

$$S_s\frac{\partial h}{\partial t} = K_f\frac{\partial^2 h}{\partial x^2} + F, \qquad (5)$$

where $F$ is the rate of recharge to the fracture zone per unit volume.

This model is characterized by three parameters:

$$S_s;\ K_f;\ F,$$

where $S_s$ is obtained from the pump test. $K_f$ and $F$ are the parameters inferred from the model.

### 3.4. One-Dimensional Groundwater Flow Model

As suggested by Roeloffs (1998) and Brodsky et al. (2003), the earthquake-induced water-level decrease can be either sudden or gradual depending on the distances from the well to the localized sources of pore pressure change. The range of the pore pressure changes can also be estimated by a simplified one-dimensional model (Fig. 3d) (Roeloffs 1996). Also, we can estimate the aquifer length by the model proposed by Wang et al. (2004c). In Wang et al. (2004c)'s model, the aquifer length could be determined by fitting the logarithm of the postseismic residuals of the groundwater level, $h$, against time, $t$, i.e.

$$\log h = a - bt, \tag{6}$$

where $a$ and $b$ are the constants for the linear fit. For sufficiently long time after the earthquake such that $t \geq \frac{\pi^2 D}{4L^2}$, we obtain the following relation (Wang et al. 2004c):

$$\frac{\partial \log h}{\partial t} \approx -\frac{\pi^2 D}{4L^2}. \tag{7}$$

This expression is independent of the spatial distribution of the coseismic source of sink of water and the location of measurement. The left-hand side of the expression is given by the slope of the field-based empirical between the hydraulic head and the time for the post-seismic change (Wang et al. 2004c). Accordingly, we can calculate the aquifer length by the fitting slope and the hydraulic diffusivity ($D$).

There are many environmental factors that may affect the records of the temporal changes, such as precipitation, tides and human activities. We should select those water-level records that are free from the environmental disturbances for study. We select a segment of the groundwater-level record and determine a post-seismic equilibrium level from the record. This is subtracted from the raw record. The difference is the post-seismic time history of the earthquake-induced groundwater-level change (i.e., the residuals). In Fig. 4, We plot the logarithm of the post-seismic residuals of the groundwater level, h,

against time, t. The coefficient $a$ is 0.07 s and $b$ is $1.72 \times 10^{-6}$ s$^{-1}$. In calculating the aquifer length, we use the hydraulic diffusivity ($D = K/S_s$ with $S_s = 1.7 \times 10^{-5}$ 1 m$^{-1}$ and $K = 0.01$ m day$^{-1}$, $D = 0.0068$ m$^2$ s$^{-1}$) and obtain the aquifer length of 98 m.

We use the solution given by Roeloffs (1998) to compute the expected time history for decay of a localized pressure drop occurring instantaneously in a one-dimension aquifer with no-flow and discharge boundaries separated by a distance $L$. For the MP well, we assume that the direction of groundwater flow in the well–aquifer system is from the right, corresponding to $X = 0$, and to the left, corresponding to $X = L$. The well is located at $X = X_w$, and the pore pressure changes in the fault zone at $X = X_0$ when the earthquake occurs. Assume that every instantaneous change occurs at the same location over the distance $|X - X_0| < \varepsilon$. An analytical solution for the change in the hydraulic head in the well, $h_p(x, t)$, is:

$$h_p(x, t) = h_0 \sum_{n=0}^{\infty} \lambda_n \cos[(2n + 1)\pi x_w/2L] \\ \times \exp[-(2n + 1)^2 \pi^2 ct/4L^2], \tag{8}$$

where

$$\lambda_m = \frac{8}{\pi(2m + 1)} \times \cos[(2m + 1)\pi x_0/2L] \sin[(2m + 1)\pi\varepsilon/2L]. \tag{9}$$

Here, $h_0$ indicates the water level at the time of the earthquake and it is the initial water level. $X_0$ is the

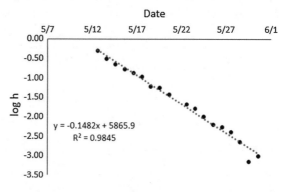

Figure 4

The logarithm of water level varying with time after earthquake

Table 1

*Model parameters for the enhanced permeability model*

|  | $h_0^a$ | $S_s^a$ | $K_f$ | $\alpha$ |
|---|---|---|---|---|
| Enhanced permeability model | 1.144 | $1.7 \times 10^{-5}$ | 0.16 | 0.07 |

[a] Fixed to this value

Table 2

*Model parameters for the transient permeability model*

|  | $h_0^a$ | $S_s^a$ | $K_1$ | $K_2$ | $\beta$ | $\lambda$ |
|---|---|---|---|---|---|---|
| Transient permeability model | 1.144 | $1.7 \times 10^{-5}$ | 0.005 | 0.011 | 0.04 | 0.04 |

[a] Fixed to this value

Table 3

*Model parameters for the increased head model*

|  | $S_s^a$ | $K_f$ | $F$ |
|---|---|---|---|
| Increased head model | $1.7 \times 10^{-5}$ | 0.015 | $2 \times 10^{-6}$ |

[a] Fixed to this value

position where the pore pressure of the fault zone changes, $\varepsilon$ is the range area of pore pressure change, $c$ is the hydraulic diffusivity and $L$ is the length of the aquifer. $t$ denotes time, and the time required for the water level to fall depends upon the distance from the

well to the nearest edge of the area of disturbed pressure (Tables 1, 2, 3).

### 3.5. Model Results

We obtain the results by fitting the Eqs. (1), (4) and (5) and use the misfit as the fitting error of the models:

$$E = \left[ \frac{1}{N} \sum_{i=1}^{N} (Q\text{measure} - Q\text{model})^2 \right]^{1/2}, \quad (10)$$

where $N$ is the number of simulated water-level data points.

The results of the four models are listed in Tables 4 and 5. We find that the hydraulic conductivity of the fault zone following the earthquake is 0.016, 0.011 and 0.015 m day$^{-1}$ for the top three models. The fitting errors of the three models are 0.04, 0.03 and 0.04. The results are similar, so all the three models fit the data well (Fig. 5). In the fourth model, we use the nonlinear least squares Marquardt–Levenberg algorithm to fit the water-level change of the MP well. Four parameters, $X_0/L$, $\varepsilon/L$, $X_w/L$ and $ct/L^2$, and the fitting errors are obtained and listed in Table 5. Using the aquifer length that was obtained from above, the following parameters, $X_0 = 78$ m, $\varepsilon = 76$ m and $X_w = 49$ m, can be obtained. The well is located at a distance from the right boundary of 49 m, and the location of the pore pressure change is 78 m from the right boundary. Thus, the location of

Table 4

*Hydraulic conductivity $K_f$ (m day$^{-1}$) inferred from the three models*

| Enhanced permeability model | Error | Transient permeability model | | Error | Increased head model | Error |
|---|---|---|---|---|---|---|
|  |  | Pre-earthquake | Post-earthquake |  |  |  |
| $0.016 \pm 0.0011$ | 0.04 | $0.005 \pm 0.0007$ | $0.011 \pm 0.0013$ | 0.03 | $0.015 \pm 0.0014$ | 0.04 |

Table 5

*The fitting results of the MP well*

|  | $X_0/L$ | $\varepsilon/L$ | $X_w/L$ | $c/L^2$ | Error |
|---|---|---|---|---|---|
| The MP well | $0.8 \pm 0.093$ | $0.78 \pm 0.15$ | $0.5 \pm 0.047$ | $0.0213 \pm 0.0058$ | 0.04 |

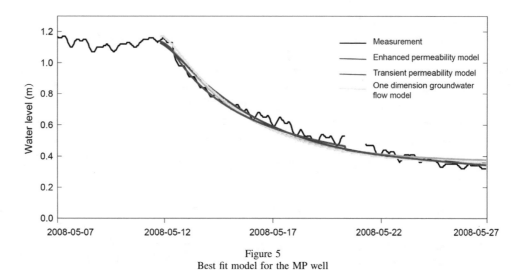

Figure 5
Best fit model for the MP well

pore pressure is 29 m away from the well, and the range area of instantaneous pore pressure change is 76 m.

## 4. Discussion

### 4.1. Mechanism of the Coseismic Groundwater-Level Changes

In the previous section, we used four models to explain the earthquake-induced groundwater flow in the MP well. All four models show good fit to the coseismic and post-seismic groundwater changes. However, the mechanisms of the four models are different.

The major difference between the increased head model and the two permeability change models is whether the permeability of the fault zone changed or not. Here, we apply baseflow recession analysis to identify whether the aquifer properties changed following the earthquake. According to (Manga 2001), the changes of hydraulic diffusivity before and after the earthquake could be identified by use of the recession constant ($aD$):

$$\frac{d \log h}{dt} = -aD. \qquad (11)$$

Here, $h$ is the groundwater level, the constant $a$ depends on the geometric properties of the groundwater system and $D$ is hydraulic diffusivity. Recession

constants will be similar if hydraulic diffusivity is unchanged before and after the earthquake. For the MP well, the pre- and post-seismic recession constants ($aD$) are $4.2 \times 10^{-3}$ $day^{-1}$ and $2.9 \times 10^{-2}$ $day^{-1}$, respectively, indicating changes of hydraulic diffusivity. Thus, an increased head model does not apply in this case.

The enhanced permeability model simulates the hydraulic conductivity in the fault zone by increasing in value permanently after the earthquake. The reason for the permanent change may be that the earthquake created new faults, thus resulting in the increase of discharge from the fault zone and a decrease in the well. However, the tidal response analysis of the MP well by Shi et al. (2013) showed a coseismic sudden increase and then a gradual recovery to the pre-seismic permeability. Thus, we exclude this model in our study.

The transient permeability model assumes that the permeability in the fault zone increased suddenly and then decreased exponentially with time, consistent with the tidal response result by Shi et al. (2013). In addition, the tidal response suggested that the hydraulic conductivity increased from 0.0046 m day$^{-1}$ before the earthquake to 0.008 m day$^{-1}$ after the earthquake. In our calculation, the hydraulic conductivity was 0.005 m day$^{-1}$ before the earthquake and increased to 0.011 m day$^{-1}$ after the earthquake. The difference of post-earthquake hydraulic conductivity between the tidal response and the model inversion may be caused by the time resolution of the tidal response method. Since the hydraulic conductivity calculated by the tidal response method in Shi et al. (2013) has a sliding window of 10 days and the post-seismic hydraulic conductivity recovered rapidly after the earthquake, the tidal response may miss the maximum changes of hydraulic conductivity. Thus, we consider the hydraulic conductivity inferred by these two methods to be consistent and we favor the transient permeability to explain the coseismic groundwater-level changes in the MP well.

The increased head model suggests that the hydraulic conductivity of the fault zone was unchanged after the earthquake, but there is increased recharge/discharge to the well from the surrounding regions. The model can also be fitted with the observed water-level data, and the hydraulic conductivity of the model inversion is $0.015$ m day$^{-1}$, which is similar to that of model 1. However, this is inconsistent with the tidal response result which showed permeability changes in the fault zone following the earthquake.

The fourth model indicated that the location of localized sources of pore pressure changes is in the down-gradient, 29 m away from of the well. As suggested by Brodsky et al. (2003), the permeability enhancement caused by the removal of the hydraulic barrier will lead to changes in the fluid pressure. The pressure response will either show steep or gradual changes depending on the distance from the barrier to the well. The location of pore pressure change (29 m) inferred from model 4 also agrees with the gradual coseismic water-level response. Thus, it is also in agreement with the conclusion derived from the transient permeability model.

### 4.2. Volume of Discharged Groundwater in the Fault Zone by the Earthquake

We now evaluate how much groundwater discharged following the earthquake in the well–aquifer system. To calculate the volume of groundwater discharge, we first calculated the unit volume of coseismic discharge, $\Delta Q$, in the fracture zone by the Theis equations (Roeloffs 1998):

$$\Delta h = \frac{\Delta Q}{4\pi H K} \int_{\frac{r^2 S_s}{4Kt}}^{\infty} \frac{e^{-y}}{y} \, dy, \qquad (12)$$

where $\Delta h$ is the water-level change and $H$ the thickness of the aquifer, $S_s$ the specific storage, $K$ the hydraulic conductivity, $r$ the distance between the calculated point and the well and $t$ the time of the water-level change after the earthquake. Before fitting the model, the groundwater-level data should be normalized by subtracting the groundwater level at the time of the earthquake.

Nonlinear least squares Marquardt–Levenberg yields a best fit of Eq. (12) to the first 20 days when $\Delta Q/4\pi H K = -0.2346 \pm 0.0198$ and $r^2 S_s/4K = 0.2914 \pm 0.0352$ (Fig. 6). We can further get the unit volume of coseismic discharge $\Delta Q$ of $0.6$ m$^3$ day$^{-1}$ with $H = 30$ m and $K = 0.01$ m day$^{-1}$. The total discharge of groundwater during the 20 days can be further constrained by the solution provided by Wang et al. (2004b):

$$Q_{ex} = \frac{2DVQ_0}{L^2(L'/L)} \sum_{r=1}^{\infty} (-1)^{r-1} \sin\left[\frac{(2r-1)\pi L'}{2L}\right]$$
$$\times \exp\left[\frac{(2r-1)^2\pi^2 D}{4L^2}t\right]. \qquad (13)$$

We calculate the excess coseismic discharge ($Q_{ex}$) by substituting the hydraulic diffusivity ($D = 0.0068 \pm 0.0008$ m$^2$ s$^{-1}$), aquifer length ($L$), the distance from the right boundary to the localized pore pressure changes ($L' = 78 \pm 10$ m), the amount

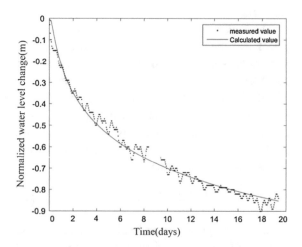

Figure 6
Use of the Theis equations to fit the change of groundwater level

of water per unit volume in the fault zone ($Q_0 = 0.6 \pm 0.08$ m$^3$) and the width of the aquifer. All of the parameters except for the width of the aquifer are obtained from the best fit of the water level by different models. Here, the aquifer width is assumed to be $22.5 \pm 2.5$ m. The coseismic discharge is then $310 \pm 90$ m$^3$. This amount of groundwater discharge has few effects on the water resource.

## 5. Conclusions

We discussed the mechanism of coseismic groundwater-level changes in the MP well following the Wenchuan earthquake. We employed four fault zone models to fit the coseismic water-level response. All of the four models fit the observation data well. By combining results from tidal response and baseflow recession analyses, we favored the transient permeability model as the plausible mechanism to explain the coseismic water-level change in the MP well. Furthermore, we obtained the location of the localized sources of pore pressure changes in the down-gradient, which is 29 m away from the well, in agreement with the conclusion derived from the transient permeability model. Finally, coseismic discharge from the MP well–aquifer system during the 20 days following earthquakes is calculated, totaling that $310 \pm 90$ m$^3$ water is discharged from the Changmutuo fault zone. Future work such as numerical modeling of the different fault zone models would be useful in determining the coseismic response mechanism. Also, more coseismic response records following other earthquakes will also be helpful in clarifying the different mechanism.

## Acknowledgements

This work was supported by the National Natural Science Foundation of China (U1602233, 41602266), the Fundamental Research Funds for the Central Universities (2652015316) and the Beijing Talents Funds (2016000020124G110). We thank guest editor Michael Manga and the two anonymous reviewers for the helpful comments.

REFERENCES

Bosl, W., & Nur, A. (2002). Aftershocks and pore fluid diffusion following the 1992 Landers earthquake. *Journal of Geophysical Research, 107*(B12), 2366.

Brodsky, E. E., Roeloffs, E., Woodcock, D., Gall, I., & Manga, M. (2003). A mechanism for sustained groundwater pressure changes induced by distant earthquakes. *Journal of Geophysical Research, 108*(B8), 2390.

Carrigan, C. R., King, G. C. P., Barr, G. E., & Bixler, N. E. (1991). Potential for water-table excursions induced by seismic events at Yucca Mountain, Nevada. *Geology, 19*(12), 1157–1160.

Che, Y., Yu, J., Liu, W., Yi, L., Xu, F., Li, J., & Sun, T. (2002). Arrangement of well network and establishment of observation well at Three Gorges of the Yangtze river (in Chinese). *Seismology and Geology, 24*, 423–431.

Elkhoury, J. E., Brodsky, E. E., & Agnew, D. C. (2006). Seismic waves increase permeability. *Nature, 441*(29), 1135–1138.

Ge, S., & Stover, S. C. (2000). Hydrodynamic response to strike- and dip-slip faulting in a half-space. *Journal of Geophysical Research, 105*(B11), 25513–25524.

Geballe, Z. M., Wang, C. Y., & Manga, M. (2011). A permeability change model for water-level changes triggered by teleseismic waves. *Geofluids, 11*(3), 302–308.

Gorokhovich, Y. (2005). Abandonment of Minoan palaces on Crete in relation to the earthquake induced changes in groundwater supply. *Journal of Archaeological Science, 32*(2), 217–222.

Gorokhovich, Y., & Fleeger, G. (2007). Pymatuning earthquake in Pennsylvania and Late Minoan crisis on Crete. *Water Science and Technology Water Supply, 7*(1), 245–251.

Lai, G., Jiang, C., Han, L., Sheng, S., & Ma, Y. (2016). Co-seismic water level changes in response to multiple large earthquakes at the LGH well in Sichuan, China. *Tectonophysics, 679*, 211–217.

Liao, X., Wang, C. Y., & Liu, C. P. (2015). Disruption of groundwater systems by earthquakes. *Geophysical Research Letters, 42*(22), 9758–9763.

Manga, M. (2001). Origin of postseismic streamflow changes inferred from baseflow recession and magnitude-distance relations. *Geophysical Research Letters, 28*(10), 2133–2136.

Manga, M., Brodsky, E. E., & Boone, M. (2003). Response of streamflow to multiple earthquakes. *Geophysical Research Letters, 30*(5), 1214.

Manga, M., & Rowland, J. C. (2009). Response of Alum Rock springs to the October 30, 2007 Alum Rock earthquake and implications for the origin of increased discharge after earthquakes. *Geofluids, 9*(3), 237–250.

Manga, M., et al. (2012). Changes in permeability caused by transient stresses: Field observations, experiments, and mechanisms. *Reviews of Geophysics, 50*(2), RG2004.

Muir-Wood, R., & King, G. C. (1993). Hydrological signatures of earthquake strain. *Journal of Geophysical Research, 98*(B12), 22035–22068.

Roeloffs, E. A. (1996). Poroelastic techniques in the study of earthquake related hydrologic phenomena. *Advances in Geophysics, 37*, 135–195.

Roeloffs, E. A. (1998). Presistent water level changes in a well near Parkfield, California, due to local and distant earthquakes. *Journal of Geophysical Research, 103*(B1), 868–889.

Roeloffs, E. A., et al. (2003). Water-level changes induced by local and distant earthquakes at Long Valley caldera, California.

*Journal of Volcanology and Geothermal Research, 127,* 269–303.

Rojstaczer, S., Wolf, S., & Michel, R. (1995). Permeability enhancement in the shallow crust as a cause of earthquake-induced hydrological changes. *Nature, 373*(19), 237–239.

Shi, Z., & Wang, G. (2014). Hydrological response to multiple large distant earthquakes in the Mile well, China. *Journal of Geophysical Research, 119*(11), 2448–2459.

Shi, Z., & Wang, G. (2015). Sustained groundwater level changes and permeability variation in a fault zone following the 12 May 2008, $M_w$ 79 Wenchuan earthquake. *Hydrological Processes, 29*(12), 2659–2667.

Shi, Z., & Wang, G. (2016). Aquifers switched from confined to semiconfined by earthquakes. *Geophysical Research Letters, 43*(21), 11166–11172.

Shi, Z., Wang, G., Manga, M., & Wang, C.-Y. (2015a). Mechanism of co-seismic water level change following four great earthquakes—insights from co-seismic responses throughout the Chinese mainland. *Earth and Planetary Science Letters, 430,* 66–74.

Shi, Z., Wang, G., Manga, M., & Wang, C. Y. (2015b). Continental-scale water-level response to a large earthquake. *Geofluids, 15*(1–2), 310–320.

Shi, Z., Wang, G., Wang, C.-Y., Manga, M., & Liu, C. (2014). Comparison of hydrological responses to the Wenchuan and Lushan earthquakes. *Earth and Planetary Science Letters, 391,* 193–200.

Shi, Z., et al. (2013). Co-seismic response of groundwater level in the Three Gorges well network and its relationship to aquifer parameters. *Chinese Science Bulletin, 58*(25), 3080–3087.

Sun, X., Wang, G., Yang, X., 2015. Coseismic response of water level in Changping well, China, to the Mw 9.0 Tohoku earthquake. J. Hydrol.

Tokunaga, T. (1999). Modeling of earthquake-induced hydrological changes and possible permeability enhancement due to the 17 January 1995 Kobe Earthquake, Japan. *Journal of Hydrology, 223*(3), 221–229.

Wang, C.-Y., Cheng, L. H., Chin, C. V., & Yu, S. B. (2001). Coseismic hydrologic response of an alluvial fan to the 1999 Chi-Chi earthquake, Taiwan. *Geology, 29*(9), 831–834.

Wang, C.-Y., Liao, X., Wang, L.-P., Wang, C.-H., & Manga, M. (2016). Large earthquakes create vertical permeability by breaching aquitards. *Water Resources Research, 52*(8), 5923–5937.

Wang, C.-Y., Manga, M., Dreger, D., & Wong, A. (2004a). Streamflow increase due to rupturing of hydrothermal reservoirs: Evidence from the 2003 San Simeon, California, Earthquake. *Geophysical Research Letters, 31,* L10502 (1–5).

Wang, C.-Y., Manga, M., Wang, C.-H., & Chen, C.-H. (2012). Transient change in groundwater temperature after earthquakes. *Geology, 40*(2), 119–122.

Wang, C. Y., Wang, C. H., & Kuo, C. H. (2004b). Temporal change in groundwater level following the 1999 ($M_w = 7.5$) Chi-Chi earthquake, Taiwan. *Geofluids, 4*(3), 210–220.

Wang, C.-Y., Wang, C.-H., & Manga, M. (2004c). Coseismic release of water from mountains evidence from the 1999 ($M_w = 7.5$) Chi-Chi, Taiwan, earthquake. *Geology, 32*(9), 769–772.

Xue, L., et al. (2013). Continuous Permeability Measurements Record Healing Inside the Wenchuan Earthquake Fault Zone. *Science, 340*(6140), 1555–1559.

Yan, R., Wang, G., & Shi, Z. (2016). Sensitivity of hydraulic properties to dynamic strain within a fault damage zone. *Journal of Hydrology, 543,* 721–728.

Zhang, L., Yang, D., Liu, Y., Che, Y., & Qin, D. (2014). Impact of impoundment on groundwater seepage in the Three Gorges Dam in China based on CFCs and stable isotopes. *Environmental Earth Sciences, 72*(11), 4491–4500.

(Received March 2, 2017, revised July 30, 2017, accepted August 1, 2017)

Pure Appl. Geophys.
© 2018 Springer International Publishing AG, part of Springer Nature
https://doi.org/10.1007/s00024-018-1925-7

**▌Pure and Applied Geophysics**

# Analysis and interpretation of earthquake-related groundwater response and ground deformation: a case study of May 2006 seismic sequence in the Mexicali Valley, Baja California, Mexico

OLGA SARYCHIKHINA,[1] ⓘ EWA GLOWACKA,[1] ROGELIO VÁZQUEZ GONZÁLEZ,[1] and MARIO FUENTES ARREAZOLA[1]

*Abstract*—Anomalous groundwater level and temperature changes are compared with ground deformation recorded before and after an earthquake of $M_W$ 5.4 and its foreshocks and aftershocks that occurred during 22–28 May 2006 in the Mexicali Valley, Baja California, Mexico. The coseismic groundwater level changes could be attributed to static volumetric strain changes caused by the mainshock, except for one well, where the groundwater level change may have been affected also by a triggered slip event at a nearby fault. Some of the coseismic temperature changes were attributed to increased convection and mixing of groundwater by seismic shaking. Modeling of groundwater level records allowed the estimation of hydraulic diffusivity. The observed ground tilt and groundwater level anomalies in the area close to the source fault before and after the mainshock and before the aftershocks occurrence are explainable by the dilatancy–diffusion theory, or possibly by assuming the occurrence of a slow slip events and/or fault permeability changes.

**Key words:** Groundwater response, ground deformation, seismic sequence, piezometer, tilt, seismic cycle.

## 1. Introduction

It is known that the fluids in the crust are very sensitive to crust strain and solid deformation. The sensitiveness of the groundwater to tectonic deformation has been observed in many cases in the form of pre-, co- and postseismic changes in physicochemical groundwater conditions (well-water level or discharge, temperature, radon concentration, etc.)

**Electronic supplementary material** The online version of this article (https://doi.org/10.1007/s00024-018-1925-7) contains supplementary material, which is available to authorized users.

[1] División de Ciencias de la Tierra, CICESE, Carretera Ensenada-Tijuana # 3918, Zona Playitas, C. P. 22860 Ensenada, Baja California, Mexico. E-mail: osarytch@cicese.mx; osarytch@yahoo.com

(e.g., Wakita 1975; Wakita et al. 1991; Quilty and Roeloffs 1997; Roeloffs 1996, 1998; Kitagawa et al. 1996; Kitagawa and Koizumi, 2000; Koizumi et al. 2004; Sil 2006; Sil and Freymuller 2006; Luo et al. 2011; Wang et al. 2012). Among them, coseismic and postseismic groundwater level changes are the most commonly reported phenomenon. Several mechanisms have been proposed to explain the hydrological phenomena related to earthquakes. Comprehensive overviews are given in Roeloffs (1996), Montgomery and Manga (2003) and Wang and Manga (2010). The behavior of groundwater responses to earthquakes is influenced by factors such as the magnitude and depth of the earthquake, distance from the epicenter and the hydrogeological characteristics around the monitoring wells. Coseismic changes within a few source dimensions of an earthquake faulting area (near-field) have often been interpreted as the poroelastic response to coseismic changes in static strain (Wakita 1975; Muir-Wood and King 1993; Quilty and Roeloffs 1997; Grecksch et al. 1999). Fracture clearing and permeability changes due to dynamic strain (seismic waves passage) have been commonly used to explain the coseismic groundwater response in intermediate and far-field (Brodsky et al. 2003; Matsumoto and Roeloffs 2003; Sil and Freymuller 2006; Wang and Chia 2008; Wang and Manga 2010). The postseismic changes are often attributed to the pressure diffusion caused by a pressure gradient in the aquifer (Wang and Manga 2010).

Precursory changes in groundwater are also reported (Roeloffs 1988; Igarashi et al. 1995; Tsunogai and Wakita 1995; Koizumi et al. 2004; Orihara et al. 2014), and they are probably related to changes in crustal strain and the generation of microfractures prior to earthquake rupture.

Documentation and analysis of groundwater responses to earthquakes (including pre-, co- and postseismic responses) are crucial for the understanding of the hydrogeological processes associated with earthquakes and the seismic cycle. In this paper, we report and interpret the hydrological effects of the May 22–28, 2006 seismic sequence that occurred in the Mexicali Valley, Baja California, Mexico. This seismicity developed in the vicinity of the Cerro Prieto Geothermal Field, which lies within the Cerro Prieto pull-apart basin, 30 km to the southeast of Mexicali city.

The first event of May 2006 seismic sequence, which was recorded by regional seismic network, occurred on 22 May at 19:57 h (UTC) and had $M_L$ 2.5. The more intense activity began with an $Mw$ 4.2 earthquake on May 24 at 04:19 h (UTC), just 1 min and 20 s before the strongest event of the sequence which reached a moment magnitude $Mw$ of 5.4. The main event was followed by a series of moderate- and low-magnitude aftershocks (Lira 2006; Munguía et al. 2009; Sarychikhina et al. 2009). In Fig. 1, the location of the events with larger moment magnitude ($Mw \geq 3.5$) is shown. The mainshock event was felt at regional distances (Mexicali–Imperial Valley region) and produced a surface rupture over more than 5 km, with maximum vertical displacement up to 30 cm and horizontal motion less than 4 cm (Suárez-Vidal et al. 2007); it also caused minor damage to local infrastructure. The mainshock was related to the Morelia fault which is a SE-dipping normal fault that trends obliquely to the major faults in the area (Fig. 1).

The May 2006 seismic sequence was recorded by local (strong motion) and regional seismic networks, as well as by continuously recording geotechnical instruments (crackmeters and tiltmeters) and well piezometers. The associated deformation was also measured by leveling profile, and synthetic aperture radar interferometry. Sarychikhina et al. (2009) estimated the mainshock source parameters using forward modeling of surface deformation data and static volume strain change (inferred from coseismic changes in groundwater level). The preferred fault model has a strike, rake, and dip of 48°, 89° and 45°,

respectively, and has a length of 5.2 km, width of 6.7 km, and 34 cm of uniform slip. Surface projection of the estimated source fault plane is shown in Fig. 1.

Here, we present and describe records of hydrological changes and ground deformation (displacement and tilting) caused by the May 2006 seismic sequence. We model the groundwater level records quantifying the coseismic and postseismic responses and estimate the hydraulic diffusivity parameter. In addition, we investigate the relationship between ground deformation and hydrological response of the groundwater level.

Note that here we were focused only on the short-term variations (1-month period centered on the time of mainshock occurrence). Only the anomalies observed in the records of two or more instruments were analyzed.

## 2. Groundwater Response

### 2.1. Groundwater Records

The groundwater response to the May 2006 seismic sequence was captured by the Mexicali Valley Piezometric Network (MVPN). The MVPN was established in 2003 with the purpose of monitoring the water level in the shallow, unconfined or partially confined, Mexicali Valley aquifer. The wells of MVPN are drilled in unconsolidated delta deposits interbedded with alluvial sediments (Vázquez González et al. 1998a). Most wells are around 150 m deep, except PZ-1 and II-9 which are 500 m and 15 m deep, respectively. In May 2006, the MVPN consisted of six piezometric wells instrumented by pressure transducers of continuous recording (Solinst levelogger) that provided data on groundwater level and temperature with 5-min sampling interval (Vázquez González et al. 2005; Fuentes Arreazola and Vázquez González 2016a, b). Figure 1 shows the location of these piezometric wells. PZ-1 well was also instrumented by barometer (barologger) whose data were used to eliminate the atmospheric influence from the groundwater level data.

The temporal evolution of groundwater level during the May 2006 seismic sequence is shown in

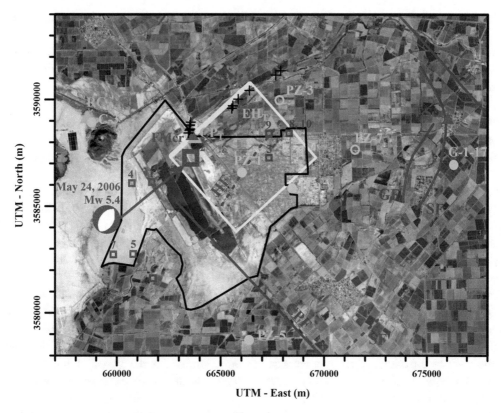

Figure 1
Location of the 2006 Mexicali Valley seismic sequence in northeastern Baja California, Mexico. Red squares show the earthquake (Mw ≥ 3.5) locations, determined by Munguía et al. (2009). The number denotes the order of earthquake occurrence. The focal mechanism (Global CMT) for the May 24, 2006 mainshock event is shown. The surface projection of the estimated by Sarychikhina et al. (2009) source fault plane is indicated by a yellow rectangle (upper edge in bold). The black crosses mark the location of sites where the surface faulting was observed by Lira (2006) and Suárez-Vidal et al. (2007). The magenta triangles indicate the location of the vertical crackmeters at the Morelia fault ($FM_{cr}$) and the Saltillo fault ($ES_{cr}$). The green triangles indicate the tiltmeter location ($RC_T$, $CP_T$ and $EH_T$). Filled and empty cyan circles indicate wells showing coseismic water level rise and drop, respectively. Thick solid red lines are known surface traces of tectonic faults: Cerro Prieto fault (CPF), Imperial fault (IF), Saltillo fault (SF), Morelia fault (MF), and Guerrero fault (GF). Dotted red lines are proposed surface fault traces based on mapped fissure zones from González et al. (1998); Glowacka et al. (2010); Lira (2006) and Suárez-Vidal et al. (2007), and well data (as the case of HF = H fault) from Lippmann et al. (1984). Black thick line frames the limits of the Cerro Prieto geothermal field. The Google Earth image is used as background

Fig. 2. The graphics cover the 1-month period from May 9 to June 8, 2006 (15 days before and 15 days after the May 24, 2006 $M$w 5.4 mainshock event of the seismic sequence). The close-up to pre-main-shock, 15-day period, piezometeric well records are presented in the electronic supplement (Figure E1).

On the records of PZ-1 and PZ-3 piezometric wells, some abnormal changes of groundwater level and temperature, which started several days before the mainshock occurrence (7 and 10 days, respectively), and before the 22–28 May 2006 seismic sequence start, have been identified. The detailed description of these changes and discussion of their possible causes are presented in the chapter 4. Additionally, groundwater level at all analyzed wells presented a low-amplitude gradual drop ($\sim$ 1–3 cm at 4–8 h) on May 22, from 9 to 5 h before the occurrence of first recorded by regional seismic network foreshock ($M_L$ 2.5). These groundwater level changes are probably related to the start of the May 22–28 seismic sequence, which probably began with low-magnitude earthquake occurrence ($M < 2.0$), not reported by regional seismic network or with aseismic slip event.

The first earthquake of the May 2006 seismic sequence which caused important changes in the groundwater level was the May 24, 2006 $M$w 5.4 mainshock event. The first measurement after the mainshock on May 24 took place about 4 min 30 s after the event (4:25:00, UTC). All six wells recorded significant, up to 6.7 m in amplitude (well PZ-1), step-like groundwater level change at the time of the mainshock event (Fig. 2, Table 1). The groundwater level change was largest at wells closest to the surface rupture. The other wells showed changes on the range of 1–55 cm (Table 1). Well II-9, which is not instrumented, showed changes in flow after the mainshock event. From the six wells which responded to the $M$w 5.4 mainshock event, two experienced a water-level drop and four experienced a water-level rise (Fig. 2).

After the May 24, 2006 mainshock event, the elevated or declined groundwater level began to recover gradually (with exception of PZ-7 well) as a result of the pressure diffusion caused by a pressure gradient in the aquifer. This groundwater level recovery causes the recharge from, or discharges to, the surrounding formations. At the PZ-1, C-3 and G-1-17 wells, the groundwater level did not recover its preseismic elevation in the time span considered in this work (Fig. 2a, c–d). At the PZ-3 well, a sudden coseismic groundwater level drop was followed by the rapid rise, instead of slow gradual recovery. Within 14 h after the coseismic fall, the groundwater level rose to a level before the mainshock, and to a level approximately 40 cm above it within ~ 3 days. The similar phenomenon was observed at PD well after the 1999 $M_{\mathrm{W}}$ 7.6 Chichi earthquake by Liu et al. (2018, Fig. 9a) who associated the large and rapid postseismic groundwater level change with the postseismic crustal deformation. The groundwater level at PZ-3 well also had a coseismic step-like response to some aftershocks after which the groundwater level began to decline but did not reach the preseismic level in the time span considered here (Fig. 2b). The preseismic groundwater level appears to be recover at the PZ-5 well; however, the earthquake-related changes observed at this well were smaller than the magnitude of the non-earthquake-related water-level fluctuations (Fig. 2f). For this reason, we excluded the PZ-5 well data from the following analysis (chapter 2.2).

At the PZ-7 well, the coseismically declined groundwater level continued to fall gradually after the May 24, 2006 mainshock with intermittent rapid falls during major aftershocks (Fig. 2e). This postseismic groundwater level decline and lack of its recovery suggest the occurrence of groundwater discharge, possibly through fractures caused by seismic shaking, which reduce the pore pressure. The similar phenomenon was observed at LY2 well in response to the 1999 $M$w 7.6 Chichi earthquake and was reported by Liu et al. (2018, Fig. 2c). The total decline of groundwater level at PZ-7 well during May 2006 seismic sequence constituted approximately 14 cm.

The step-like groundwater level changes in response to $M$w 4.4 (28/05/2006 07:40:20) and the $M$w 4.6 and $M$w 3.9 (28/05/2006 11:55:23 and 11:57:47) aftershocks are also observed in the groundwater level records of the PZ-1, PZ-3 and PZ-7 wells. The groundwater level change in response to the $M$w 4.5 aftershock (27/05/2006 10:21:35) is observed in groundwater level record of PZ-1 well.

PZ-1 and PZ-3 wells also recorded groundwater temperature changes caused by mainshock event (Fig. 2). The close-up to 1-day period centered on the time of mainshock occurrence of groundwater temperature record at PZ-1 and PZ-3 is presented in the electronic supplement (Figure E2). At both wells, groundwater temperature drop by 0.1°C at PZ-3 and 0.3°C at PZ-1 was followed by a gradual rise. Coseismic water temperature changes are known to be caused by increased convection and mixing of groundwater as a result of ground shaking. Factors influencing such changes include stress condition of the relevant aquifer, temperature gradient around the well, flow direction change caused by seismic waves, and the location of water-temperature probe in the well (Wang and Manga 2010 and refs. therein). After the occurrence of earthquake, as the fluctuation of water level gradually quiets down, water temperature begins to rise.

## 2.2. Modeling of Groundwater Level Records

As discussed earlier, the six piezometric wells recorded step-like coseismic groundwater level changes caused by May 24, 2006 mainshock event followed by gradual postseismic changes.

In general, the groundwater level records analyzed here and presented in Fig. 2 are formed by three stages. Here, we used the equation of Sil (2006) to fit the groundwater time series using a combination of a linear trend (for preseismic), a step function (for coseismic) and an error function (for postseismic) of the form:

$$w(t) = I + St + CH(t - t_0) + P\mathrm{erfc}\left(\sqrt{\tau/(t - t_0)}\right)$$
(1)

where $I$ (intercept) and $S$ (slope) are the constants that describe the linear pre-earthquake trend, $H$ is the Heaviside function, $C$ is a constant (magnitude of the coseismic step), erfc is the complementary error function, $P$ is a constant (magnitude of the postseismic changes), and $\tau$ is the characteristic decay time of the groundwater level. The time of the mainshock event is $t_0 = 0$ and $w(t)$ is the observed groundwater level at any time $t$ of the time series. We used least-squares approach to fit the groundwater level data and to simultaneously estimate all parameters before and after the mainshock $M$w 5.4 event.

The Curve Fitting Tool of the Matlab software was used for fitting. Before the fitting, the obvious outliers were manually excluded from the records to reduce the errors and residuals of modeling.

For three of the five remaining wells PZ-1, C-3 and G-1-17, we selected 1 month of groundwater level record (15 days before and 15 days after the mainshock event). For PZ-3 and PZ-7, we considered the time of 15 days before the mainshock occurrence and 4 days after as the May 28, 2006 aftershocks caused important step-like groundwater level change (see Fig. 2) at these wells.

The results of the modeling and the residuals are shown in Figs. 3; the estimated parameters are presented in Table 2. Large residuals (observed records minus fit) occur over the period close to the mainshock occurrence and we attribute them to the "wellbore storage" effect. As it can be seen in Fig. 2, the coseismic groundwater level response is not instantaneous; groundwater level in the well cannot track the arbitrarily fast changes of aquifer pressure because a change of groundwater level in the well requires water to flow into or out of the well. This degradation of groundwater level response to aquifer pore pressure is often termed "wellbore storage" and depends on the characteristics of each particular well. The groundwater level changes following the May 24, 2006 earthquake show that the largest wellbore storage effect is at the location of C-3 well, where it took about 10 h for the groundwater level to reach its highest value after the earthquake. In addition to wellbore storage effect, the postseismic pore fluid diffusion due to the coseismic response of pore pressure field may be the cause.

Sarychikhina et al. (2009) suggested that the recorded coseismic May 24, 2006 groundwater level changes can be explained by the poroelastic response (Biot 1941; Rice and Cleary 1976; Kümpel 1991; Wang 1993; Roeloffs 1996). The static volume strain change, inferred from coseismic changes in groundwater level, coupled with surface deformation data was used to estimate the source parameters for the May 24, 2006 mainshock by Sarychikhina et al. (2009). The observed coseismic groundwater level changes and the static volume strain change at well locations, estimated by the modeling, were used to obtain the value of static volumetric strain sensitivity ($4.5 \times 10^{-2}$ cm/n$\varepsilon$; $1n\varepsilon = 10^{-9}$), as a best least-squares fit between these two quantities with coefficient of determination, $R^2$, of 0.994. Here, using these two quantities, the static volume strain sensitivity for each well location was calculated and is presented in Table 1. It varies from 0.4 (PZ-7) to 9.3 (G-1-17) cm/n$\varepsilon$, with an average value of 3.9 cm/n$\varepsilon$. The standard deviation of the estimated values of the static volume strain sensitivity is $\pm 2.8$ cm/n$\varepsilon$, and all values, except for two extreme values (PZ-7 and G-1-17), are within 1 standard deviation from the average.

Fuentes Arreazola and Vázquez González (2016b) and Fuentes Arreazola et al. (2018) analyzed the tide responses of the MVPN wells to estimate coefficient relating groundwater level to strain. In the study by Fuentes Arreazola and Vázquez González (2016b), a tidal analysis was done using data recorded before (April 27–May 21, 2006) and after (June 4–July 19, 2006) the May 22–28, 2006 seismic sequence. There

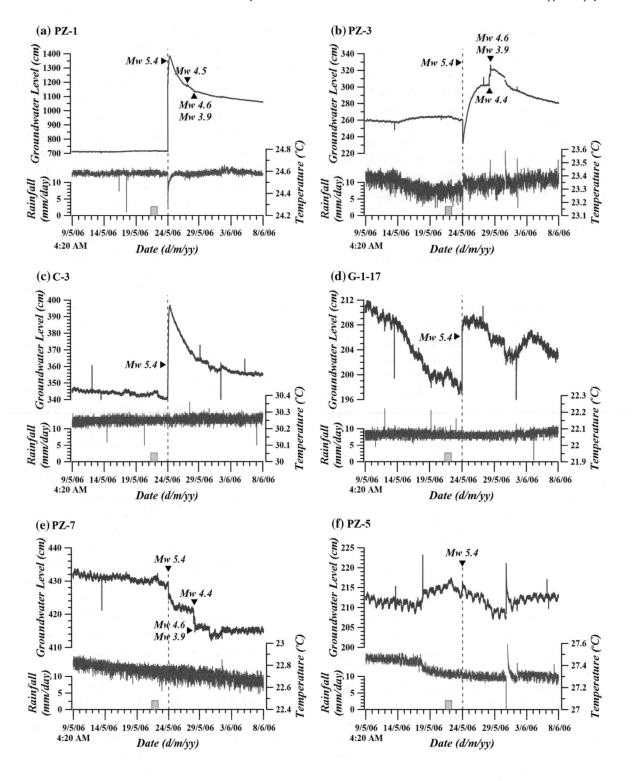

Groundwater level and temperature records showing the changes related to the May 2006 seismic sequence. The groundwater level records are corrected by atmospheric pressure. The timescale covers a range of 15 days before and after the mainshock. The sampling interval is 5 min. The significant step-like groundwater level changes caused by mainshock (Mw 5.4) and aftershocks (Mw > 3.5) are indicated. The Mw 4.5 aftershock is #6 in Fig. 1, Mw 4.4 aftershock is # 8 and the Mw 4.6 and Mw 3.9 aftershocks are # 9 and #10, respectively. Rainfall data from nearest meteorological station are also presented

was not significant difference found between the tidal strain responses before and after May 2006 seismic sequence. In the study by Fuentes Arreazola et al. (2018), a tidal analysis was performed using data recorded between May 23 and August 14, 2007 (83 days). The values of areal strain sensitivity obtained by these studies are presented in Table 1.

The strain sensitivities determined by coseismic responses in the groundwater level were compared with those determined by tidal strain. Before the comparison, the static volume strain sensitivities obtained from coseismic groundwater level responses were converted to areal strain sensitivities using Eq. (1) from Wang and Barbour (2017) and Poisson ratio of 0.25. The compared values are consistent at the majority of analyzed wells. At PZ-3 and PZ-5 wells, these values are practically identical. Some

small discrepancies of these values at PZ-1, PZ-7 and C-3 wells could be attributed to the simplicity of dislocation model used for static volume strain change estimation. The non-elastic response of groundwater level could contribute as well.

However, at G-1-17 well the strain sensitivity determined by coseismic response is almost one magnitude order larger than estimated by tidal strain. The observed amplitude of groundwater level change at this well is larger than expected from the static volume strain imposed by mainshock. It could be due to the well location near the Saltillo fault where the slip was triggered by the mainshock event. The groundwater level anomalies of the order of centimeters at G-1-17 well related to the slip events on the Saltillo fault were reported by Glowacka et al. (2011). So, the amplitude of groundwater level change at G-1-17 well could be due to combined effects (poroelastic and/or non-elastic) caused by mainshock and by triggered slip on the Saltillo fault.

The values of $\tau$ obtained from the fitting were used for the determination of the scalar apparent hydraulic diffusivity. Assuming that the source of the pore pressure changes is located around the coseismic fault plane (Sarychikhina et al. 2009), the minimum distances ($z$) to the fault plane from the well locations were obtained and the hydraulic diffusivities ($D$) were estimated using equation:

Table 1

*Observed coseismic groundwater level changes ($\Delta h$), estimated from the modeling (preferred fault model; Sarychikhina et al. (2009)) coseismic static volume strain change ($\Delta\varepsilon$), and strain sensitivity determined by coseismic responses in the groundwater level and by $M_2$ tidal strain. Vs and As mean volume and areal strain sensitivity, respectively. Negative values of static volume strain change mean contraction*

| Well | $\Delta h$ | $\Delta\varepsilon$ | Strain sensitivity (cm/n$\varepsilon$) $\times 10^{-2}$ | | | | |
|------|------|------|------|------|------|------|------|
| | | | Vs | As[a] | As ($M_2$)[b] | | As ($M_2$)[c] |
| | (cm) | (n$\varepsilon$) | | | Pre- | Post- | |
| PZ-1 | 670 | − 1.48E + 04 | 4.53 | 3.02 | | | 1.87 |
| PZ-3 | − 28 | 1.50E + 03 | 1.87 | 1.24 | 1.95 | 1.59 | 1.38 |
| C-3 | 55 | − 1.27E + 03 | 4.33 | 2.89 | 1.44 | 1.69 | 1.40 |
| G-1-17 | 10 | − 1.07E + 02 | 9.35 | 6.23 | 1.22 | 0.94 | |
| PZ-7 | − 3 | 7.08E + 02 | 0.42 | 0.28 | 1.69 | 1.24 | |
| PZ-5 | 1 | − 3.19E + 01 | 3.13 | 2.09 | 2.11 | 2.20 | |

[a]Estimated using Vs, Eq. (1) from Wang and Barbour (2017) and Poisson ratio of 0.25

[b]From Fuentes Arreazola and Vázquez González (2016)

[c]From Fuentes Arreazola et al. (2018)

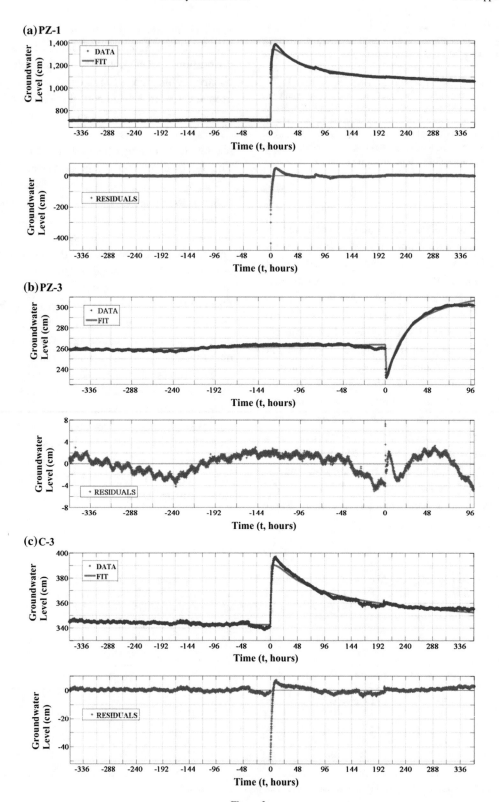

Figure 3

Modeled (Eq. 1) and observed groundwater level records during the May 2006 seismic sequence. $t_0 = 0$ is the mainshock occurrence time

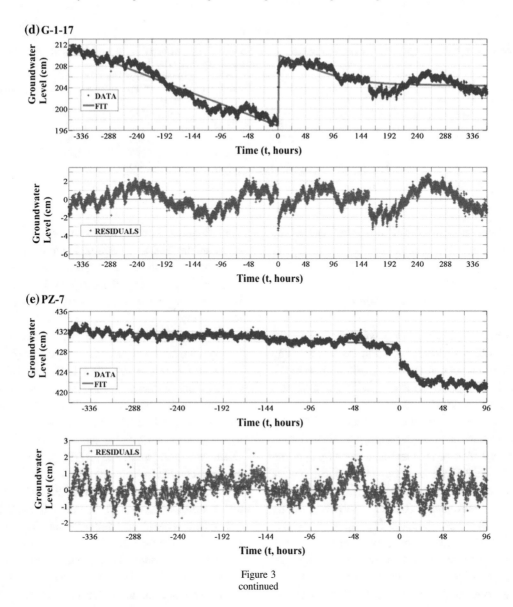

Figure 3
continued

$$\tau = z^2/4D \qquad (2)$$

The results are shown in Table 2. For four of the five analyzed wells PZ-1, PZ-3, C-3 and G-1-17, the range of $D$ values estimated here (from 7.39 to 12.8 m²/s, with an average value close to 10 m²/s) is in good agreement with the published $D$ value range. Talwani and Acree (1985) calculated the "seismic" hydraulic diffusivity estimates ranging from 0.5 to 50 m²/s and Simpson et al. (1988) calculated estimates of 1–10 m²/s. For the Mexicali Valley, Glowacka and Nava (1996) and Glowacka et al.

(2010) estimated the hydraulic diffusivity values ranging from 1 to 30 m²/s for the seismicity related to the fluid extraction (pressure decrease) in the CPGF. However, the estimated $D$ value at PZ-7 exhibits relatively high value of ~ 323 m²/s, one magnitude order higher than the other wells and the published $D$ values, suggesting that for this well the assumption that the source of the pore pressure changes is located around the coseismic fault plane is erroneous. As it was suggested earlier, the postseismic changes at this well probably reflect the local permeability

Table 2

*Coseismic (C) and postseismic (P) groundwater level changes and decay constants (τ) obtained by the modeling of the groundwater level records, and estimated hydraulic diffusivity (D)*

| Well ID | C (cm) | P (cm) | τ (h) | $R^2$ | RMSE (cm) | z (m) | Hydraulic diffusivity D (m²/s) |
|---------|--------|--------|-------|-------|-----------|-------|-------------------------------|
| PZ-1    | 621    | − 423  | 28    | 0.997 | 12.05     | 1780  | 7.73                          |
| PZ-3    | − 31   | 104    | 8     | 0.985 | 1.74      | 1130  | 11.24                         |
| C-3     | 48     | − 54   | 37    | 0.964 | 2.45      | 2600  | 12.80                         |
| G-1-17  | 13     | 74     | 422   | 0.907 | 1.07      | 6700  | 7.39                          |
| PZ-7    | − 3    | − 5    | 4     | 0.972 | 0.62      | 4500  | 323.35                        |

z is the minimum distance from well location to the coseismic fault plane. RMSE and $R^2$ values of the modeling are also presented

enhancement as a result of fracturing caused by seismic shaking.

## 3. Geotechnical Instrument Records

Geotechnical instruments have operated in the Mexicali Valley since 1996 for continuous recording of deformation phenomena. Up to 2006, the network consisted of three crackmeters and eight tiltmeters (Glowacka et al. 2007, 2008).

Crackmeters (extensometers) measure fault slip (vertical or horizontal) by recording the displacement between two piers or monuments located on opposite sides of a fault. Crackmeters of the Mexicali Valley geotechnical instruments network (REDECVAM) are 3-m long crackmeters (Geokon Vibrating Wire, model 4420) which span the fault in a plane perpendicular to it and sloping with respect to the horizontal. The crackmeters have 70-cm range and 0.1 mm precision, and operate with a 5–20-min sampling interval (Glowacka 1996; Nava and Glowacka 1999; Glowacka et al. 2010). During the time period covered by this study, the crackmeters operated with a 20-min sampling interval.

Tiltmeters are highly sensitive instruments used to measure changes in the inclination of the ground. These instruments are usually installed in boreholes to avoid spurious ground tilts produced by differential thermal expansion in near-surface materials and rainfall effects (Agnew 1986). Borehole and surface tiltmeters (Applied Geomechanics, models 722 and 711) of REDECVAM are biaxial tiltmeters, which means that they measure two components (north–south and east–west in this case) of earth surface tilt.

The tiltmeters have a measuring range of $\pm 2000$ microradians and resolution of 1 microradian minute (Glowacka et al. 2002). During the time period spanned by this study, the sampling interval of surface and boreholes tiltmeters was 2 and 4 min, respectively.

The May 2006 seismic sequence was recorded by five instruments: two vertical crackmeters ($FM_{cr}$ and $ES_{cr}$), two borehole tiltmeters ($EH_T$ and $RC_T$) and one surface tiltmeter ($CP_T$). The location of the instruments is presented in Fig. 1. The vertical crackmeters are installed at the Morelia fault ($FM_{cr}$), which was responsible for the May 24, 2006 mainshock, and at the Saltillo fault ($ES_{cr}$). The records of the geotechnical instruments spanning the May 2006 seismic sequence are presented in Fig. 4. The close-up to pre-mainshock, 15-day period, geotechnical instrument records are presented in the electronic supplement (Figure E3).

The first recorded by regional seismic network May 22 $M_L$ 2.5 foreshock caused no detectable changes in the records of geotechnical instruments. However, the geotechnical instruments installed on or close to Morelia fault ($FM_{cr}$ and $CP_T$) report fault displacements and tilt changes $\sim 6$ h before this foreshock. As it can be seen in Figure E3, since May 22 Morelia fault presented several movements of low amplitude ($\leq 1$ mm) before the mainshock event. These observed fault movements are probably related to the start of the May 22–28 seismic sequence, which probably began with low-magnitude earthquake occurrence ($M < 2.0$), not reported by regional seismic network or with aseismic slip event (as it was proposed in chapter 2.1).

Table 3

*Coseismic displacements and tilt changes observed by geotechnical instruments and predicted by the preferred fault model (Sarychikhina et al. 2009)*

| Geotechnical instrument | Observed | | Predicted | |
|---|---|---|---|---|
| | Displacement (mm) | | | |
| FM$_{cr}$ | 70 (200–250[a]) | | ~ 240 | |
| ES$_{cr}$ | 2 | | | |
| | Tilt (microrad) | | | |
| | North | East | North | East |
| EH$_T$ | − 420 | > 2000 | 4 | 50 |
| RC$_T$ | − 21 | − 40 | − 0.695 | 0.975 |
| CP$_T$ | − 170 | − 930 | 920 | − 900 |

Downward tilt to the east or the north is defined as positive

[a]Displacement after correction for installation angle and for instrument separation from the benchmark

The observed amplitudes of surface displacement and tilt offset caused by mainshock event are listed in Table 3.

The record of the FM$_{cr}$ showed that the total motion during the May 24, 2006 mainshock, after corrections for installation angle (30° from the horizontal) and instrument separation from the benchmark, is 20–25 cm, which matches the reported field observations of surface offset. The data from the FM$_{cr}$ were used by Sarychikhina et al. (2009) for the source fault parameters modeling, because this crackmeter directly reports the coseismic motion in the fault which generated the earthquake.

The record of the ES$_{cr}$ showed 2 mm of vertical motion on the Saltillo fault, probably triggered by the mainshock in Morelia fault. The ES$_{cr}$ is located ~ 13 km to the east of the epicenter. Saltillo fault is characterized by the phenomenon of stable creep and episodic fault slip which appears mainly as slip-predictable, normal, aseismic slip. However, the triggering of Saltillo fault slip by seismic events was also observed and reported by Glowacka et al. (2002).

The tilt meters report impulsive tilt variation (offset) in two components, correlated with May 24, 2006 mainshock event (Fig. 4). The observed tilt offset amplitudes are characterized by simultaneous occurrence at recording stations and values ranging between 21 (north–south component of RC$_T$) and more than 2000 microradians (east–west component of EH$_T$). The aftershocks ($Mw > 3.5$) also caused some important tilting effects detected by the EH$_T$.

The tiltmeters data were not included in the source fault modeling by Sarychikhina et al. (2009) because the shallow and surface tilt sensors are influenced not only by the surface displacement caused by the earthquake but also by motions on cracks, fractures, and on minor faults near the instrument site, as well as by instability of the benchmarks, as has been noted for other earthquakes (McHugh and Johnston 1977; Wyatt 1988; Takemoto 1995).

Using the preferred model (Sarychikhina et al. 2009) and the EDCMP software (Wang et al. 2003), the north–south and east–west tilt components were estimated and are presented in Table 3. Strong discrepancies between observed and predicted coseismic tilt variations confirm the initial supposition about different sources of influence on the coseismic tiltmeter records. The tilt values recorded by RC$_T$ can be influenced by a local fault movement triggered by the mainshock, as the RC$_T$ is installed very close to proposed trace of northern extension of the Cerro Prieto fault (Magistrale 2002; Glowacka et al. 2015). The discordance in CP$_T$ can be caused by the non-uniform strike along the trace of Morelia fault (Fig. 1), which changes its strike from ~ 48° to ~ 0° close to the epicenter and CP$_T$ location. This Morelia fault geometrical heterogeneity was not considered in the source fault model proposed by Sarychikhina et al. (2009).

Using equations published by Freund and Barnett (1976), Cohen (1996, 1997) and Glowacka et al.

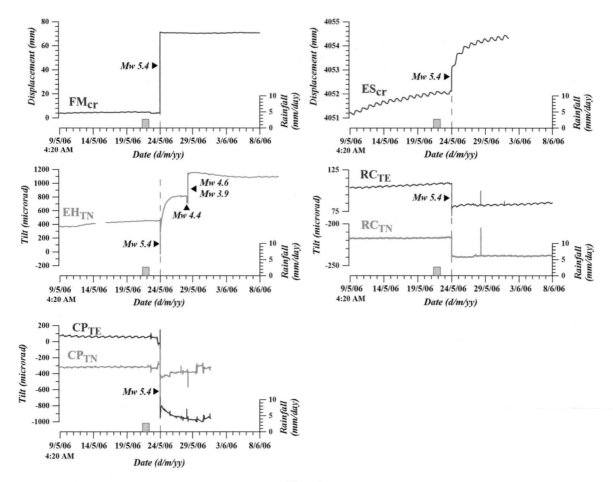

Figure 4
Geotechnical instruments records showing the surface displacement and tilt related to the May 2006 earthquakes sequence. The important permanent step-like (not impulse) changes caused by mainshock (Mw 5.4) and aftershocks (Mw > 3.5) are indicated. The Mw 4.4 aftershock is # 8 in Fig. 1 and the Mw 4.6 and Mw 3.9 aftershocks are # 9 and #10, respectively. For the tiltmeters, two components, north–south (TN) and east–west (TE), are presented (except $EH_T$ east–west component which was out of range). Downward tilt to the east or the north is defined as positive. Subscript cr means crackmeter. The timescale covers a range of 15 days before and after the mainshock. Rainfall data from nearest meteorological station are also presented

(2017) calculated tilt change on $EH_T$ location related to the coseismic slip on the Morelia fault or to the possible triggered slip on the Hidalgo fault (Fig. 1). Several combinations of fault slip, dip and depth were tested. However, no combination of the tested fault parameters produces the tilt changes similar to the ones recorded by $EH_T$. The authors concluded that the slip on the fault, or fracture, very close (few meters) to the $EH_T$ could be responsible for these recorded tilt change values. The other possibility (not modeled) is the abrupt change in the phreatic level since the $EH_T$ is installed in a place where phreatic level is very shallow.

## 4. Discussion

The records of PZ-1, PZ-3 and $EH_T$ presented important changes which started several days before the mainshock occurrence (from 12 to 7 days), and before the May 22–28, 2006 seismic sequence start.

The PZ-3 well is located close to the eastern ruptured fault end, in its hanging wall, and in the area where four aftershocks of $Mw > 3.5$ occurred (Fig. 1). Groundwater level in PZ-3 started to increase gradually 10 days before the mainshock occurrence, whereas the groundwater temperature gradually decreased (Fig. 5).

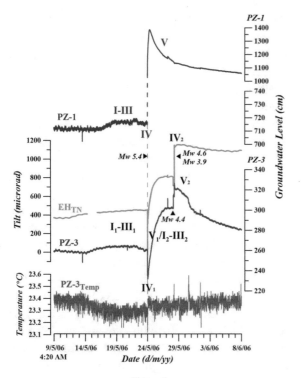

Figure 5
Groundwater level (PZ-1, PZ-3) and temperature (PZ-3$_{Temp}$), and tilt (EH$_{TN}$) records. Roman number indicates the corresponding stage of dilatancy–diffusion theory (as in Scholz et al. (1973)) identified in PZ-1 record (violet numbers) and in EH$_{TN}$ and PZ-3 records (black numbers). Two cycles were identified in EH$_{TN}$ and PZ-3 records. Subscript 2 denotes the stages of second cycle. See text for details

The EH$_T$, located $\sim 1$ km to the south of the PZ-3 well, presents a gradual tilt increase toward north (north–south tilt component) which started 12 days before the mainshock occurrence (Fig. 5).

The PZ-1 well, located in the hanging wall, in the central area of the surface projection of rupture (Fig. 1), presents a gradual increase of groundwater level which starts 7 days before the mainshock occurrence (Fig. 5). There was no detectable groundwater temperature change before the mainshock at PZ-1.

These changes apparently are not associated with rainfall, local pumping or fluid injection. The Mexicali Valley has an arid climate and low annual precipitation ($\sim 71$ mm). There are two meteorological stations in the area of study, and the nearest to PZ-1, PZ-3 and ET$_T$ station ($\sim 2$ km apart) reported some low precipitation ($\sim 2.7$ mm) 2 days before mainshock (Figs. 2, 4); the other station ($\sim 8$ km

apart) did not report any precipitation. There is no local pumping from shallow aquifer. The PZ-1 well is located in the area of geothermal extraction wells; however, there is no connection between shallow aquifer and deep ($> 2500$ m) geothermal reservoirs (Vázquez-González et al. 1998b). The wells which re-inject the geothermal brine are located 5 km to the West from PZ-1, outside the Cerro Prieto pull-apart basin. The range of re-injection depth is 500–2600 m. The borehole tiltmeter EH$_T$ (5 m depth) has some sensitivity to agriculture field irrigation activities (CONAGUA, personal communication; Electronic Supplement, Figures E4–E6). However, the PZ-3 groundwater level and temperature records do not present changes related to these activities, probably due to a greater depth (150 m) (Electronic Supplement Figure E7) nor do the PZ-1 groundwater and temperature records, as it is located at distance from the agriculture field (Electronic Supplement Figure E8). So, the observed changes are most likely reflecting the preseismic processes that occurred at the fault zone.

In what follows, the several possible scenarios, which could explain the observed pre-seismic and postseismic abnormal changes at PZ-1, PZ-3 and EH$_T$ records, are discussed.

First scenario is suggested by dilatancy–diffusion theory. Scholz et al. (1973; Fig. 3) present the predictions of anomalies in geophysical signals associated with stages of elastic loading (stage I), dilatant yield (stage II), pore pressure recovery (stage III), dynamic failure (stage IV), and postseismic relaxation (stage V) which are similar to the ones observed in EH$_T$ (north–south component, EH$_{TN}$), PZ-3 and PZ-1 records. The stages we identified are indicated in Fig. 5.

At PZ-1 well, the gradual increase of groundwater level starts 7 days before the mainshock occurrence (stage I–III). The mainshock imposes the contraction in the area of PZ-1 location inducing the groundwater level rise (IV), following the gradual postseismic groundwater level recovery (decrease; stage V).

It is important to highlight a high degree of similitude between the PZ-3 groundwater level and EH$_{TN}$ tilt records at all times, i.e., in pre-, co- and postseismic periods (Fig. 5). Two cycles can be identified in the EH$_{TN}$ and PZ-3 records. The first cycle starts before the mainshock occurrence (12 days for $EH_{TN}$ and 10 days for PZ-3), and manifests itself by means of a gradual increase of the groundwater level and gradual decrease of the groundwater temperature, caused probably by influx of water of lower temperature, and tilt increase toward north (stage $I_1$–$III_1$). The earthquake produces the groundwater level and temperature drop and tilt decrease (stage $IV_1$). After the coseismic changes, a rapid rise of tilt and groundwater level and temperature, to the level above the level before the mainshock, is observed. Since PZ-3 is located in the dilatation zone imposed by mainshock (Sarychikhina et al. 2009, Fig. 5), it could be greatly affected by postseismic water diffusion from the contraction zone. So, the coseismic response and subsequent water diffusion may cause this observed rapid groundwater level rise ($V_1$). On the other hand, PZ-3 and $EH_T$ are located in the area affected by the aftershocks activity. So, there is a possibility that these postseismic rapid groundwater level and tilt rises could be a reflection of preparation for the aftershock activity (second cycle, stage $I_2$–$III_2$), or of combined effect of these two discussed processes (stage $V_1/I_2$–$III_2$). The aftershocks occurrence (stage $IV_2$) and post-aftershocks gradual recovery (stage $V_2$) were also identified in tilt ($EH_{TN}$) and groundwater level (PZ-3) records.

The other possibility to explain the observed anomalies in Fig. 5 is the occurrence of a slip event on the Morelia fault before the mainshock and before the aftershocks. Lohman and McGuire (2007) suggested that 2005 Obsidian Buttes swarm was triggered by the shallow aseismic fault slip, and Wei et al. (2015) suggested that Brawley swarm, and particularly M 5.3 earthquake, was triggered by aseismic creep caused by fluid injection. The faults of Mexicali Valley present aseismic creep induced by the fluid extraction in the CPGF (e.g., Glowacka et al. 2010). While both components of $CP_T$, installed close to the southern part of Morelia fault, present possible creep of $\sim 250$ (east–west component, $CP_{TE}$) and $\sim 700$ (north–south component, $CP_{TN}$) microradians during 2004–2006, the $FM_{cr}$, located close to $CP_T$ ($\sim 150$ m) on the fault, did not record creep during this period (electronic supplement Figure E9).

There are creep and sporadic slip events in the Saltillo fault (e.g., Glowacka et al. 2010;

Sarychikhina et al. 2015), and tiltmeter and extensometer installed on this fault, very close to each other ($\sim 100$ m), show very similar form of anomalies related to creep and slip events (Sarychikhina et al. 2015, Fig. 6). Some of the slip events were also associated with the groundwater level anomaly of the order of centimeters (Glowacka et al. 2014). So, we cannot exclude that the anomaly observed at PZ-3, PZ-1, and $EH_{TN}$, before the start of the seismic sequence and after the mainshock event, is related to the slip events.

Third possible scenarios, which could explain the abnormal groundwater level changes and ground tilt anomalies before and after the mainshock (Fig. 5), assume the change in permeability of fault (process described, for example, by Manga et al. 2012). According to Elkhoury et al. (2006), an earthquake can increase permeability of the site by a factor as high as three. King (2018) (this volume) observed the phenomenon of groundwater level decreases before and after the strong distant earthquakes in sensitive wells in Tono Mine, central Japan. This phenomenon is explained as an effect of slow slip before the earthquake and dynamic triggering of fault permeability increase. The observations were performed in the area where an impermeable fault separates zones with differential aquifer pressures, and sensitive well is located on the side with a higher pressure.

In the area of this study, there is a difference in reservoir pressure caused by fluid extraction in the Cerro Prieto geothermal field. The lowering of pressure in the deep geothermal reservoirs (1000–3500 m) and surrounding rocks due to extraction of geothermal fluids results in rapid subsidence in the area of extraction wells and recharge zone (Grecksch et al. 1999). Sarychikhina et al. (2018) report the subsidence rate up to 20 cm/year during 1993–2014. Since the subsidence is abruptly limited by Saltillo, Imperial and Cerro Prieto faults, Grecksch et al. (1999) and Sarychikhina et al. (2011) proposed that those faults are impermeable in direction perpendicular to the fault. Although Suárez-Vidal et al. (2008) proposed that Morelia fault is a northern limit of Cerro Prieto pull-apart basin, we cannot be sure that Morelia fault is impermeable, since it forms a zone of faults (Fig. 1), and it does not constitute a sharp limit of the subsidence

(Sarychikhina et al. 2018, Figs. 9, 10 and 12). However, we can suppose that Morelia fault is impermeable on the segment close to PZ-1, PZ-3 and EH$_T$. In this case, the slip event preceding the mainshock, as proposed in the previous paragraph, can change the permeability of the fault by increasing fracturing and causing water infiltration from the area outside the subsidence bowl (area with a higher pressure) thus causing groundwater level gradual increase in wells close to the Morelia fault (PZ-1, PZ-3). The groundwater level increase preceding aftershocks could be in part a continuation of water infiltration from outside the Cerro Prieto subsidence bowl due to permeability increase caused by mainshock. Unfortunately, we do not have groundwater level data from the northern side of Morelia fault to confirm this idea. Since the EH$_T$ is probably sensitive to the pressure changes in the phreatic level, as can be seen in figure E5, pre- and postseismic anomalies of tilt can be simply explained as changes caused by pressure changes in the aquifers below the instrument.

More analysis is necessary to confirm or reject these scenarios.

## 5. Summary and Conclusions

Documentation and analysis of groundwater responses to the earthquakes allow a better understanding of their influence on shallow groundwater systems and hydrogeological properties and processes. Here, we analyzed the short-term variations (1-month period) of ground deformation (displacement and tilt) and groundwater level and temperature in relation to the May 22–28, 2006 seismic sequence in the Mexicali Valley, Baja California, México.

The mainshock (Mw 5.4) of May 24 caused important ground deformation and notable changes in the groundwater level and temperature in the Mexicali Valley. The coseismic (mainshock) groundwater level changes were explained by the poroelastic response to the earthquake's strain field by Sarychikhina et al. (2009). Here, we compared the strain sensitivities determined by coseismic responses in the groundwater level and by tidal strain. The results of this comparative analysis suggest that, for the

majority of analyzed wells, the amplitude of observed coseismic groundwater level change is primarily related to the coseismic static volume strain change imposed by the mainshock. The non-elastic response of groundwater level could contribute as well, but less so. Exception: G-1-17 well, where the amplitude of groundwater level change was probably due to combined effects caused by mainshock and by triggered slip on Saltillo fault.

The exploratory analysis and modeling of groundwater level records performed here suggest that at four of the five analyzed piezometric wells the postseismic pattern of the groundwater level changes could be explained by the gradual diffusion of pore pressure. The characteristic decay time is lower for the well located closer to the coseismic rupture. Assuming the coseismic fault plane as the source of pore pressure changes and using the characteristic decay time obtained by the modeling, the hydraulic diffusivity was estimated. The range of estimated values is in good agreement with values obtained by the other works for the same area (Glowacka and Nava 1996; Glowacka et al. 2010), except for PZ-7 well. At PZ-7, the observed postseismic groundwater level changes suggested the occurrence of groundwater discharge, possibly through fractures generated by the mainshock.

The coseismic groundwater temperature decrease at PZ-1 and PZ-3 could be explained by increased convection and mixing of groundwater as a result of ground shaking.

Three possible scenarios which could explain the observed ground tilt and groundwater level anomalies in the area close to the source fault before and after the mainshock occurrence were proposed. The first scenario includes the phenomena described by the dilatancy–diffusion theory. The possibility of a slip preceding the mainshock event and slip event preceding the aftershocks, as well the possibility of fault permeability enhancement were also proposed and discussed.

The observed phenomena seem to be important for earthquake prediction and more detailed analysis of more pre-, co-, and postseismic data is needed to obtain more credible conclusion.

Our results underline importance of continuous and long-term groundwater and deformation monitoring with the attention that both sides of faults which separate aquifers with different pressure and/or faults with creep should be instrumented.

## Acknowledgements

This research was sponsored in part by Consejo Nacional de Ciencia y Tecnología (CONACYT), project number 45997-F, and CICESE internal funds. Authors appreciate the access to seismicity data from RESNOM. Our thanks to Francisco Farfán, Guillermo Diaz de Cossio, Luis Orozco y Oscar Gálvez for help during the fieldwork and to Miguel Angel Garcia Arthur and students Adriana Perez Martinez and Antonio Garcia Hernandez for help in data processing. The authors greatly acknowledge the editor, Dr. Chi-Yu King, and one anonymous reviewer for their valuable comments and suggestions which helped to improve the manuscript.

## REFERENCES

Agnew, D. C. (1986). Strainmeters and tiltmeters. *Reviews of Geophysics, 24,* 579–624.

Biot, M. A. (1941). General theory of three-dimensional consolidation. *Journal of Applied Physics, 12*(2), 155–164.

Brodsky, E. E., Roeloffs, E., Woodcock, D., Gall, I., & Manga, M. (2003). A mechanism for sustained groundwater pressure changes induced by distant earthquakes. *Journal of Geophysical Research, 108*(B8), 2390. https://doi.org/10.1029/2002jb002321.

Cohen, S. C. (1996). Convenient formulas for determining dipslip fault parameters from geophysical observables. *Bulletin of the Seismological Society of America, 86,* 1642–1644.

Cohen, S. C. (1997). Convenient Formulas for Determining Dip-Slip Fault Parameters from Geophysical Observables (ERRATUM). *Bulletin of the Seismological Society of America, 87,* 1081.

Elkhoury, J. E., Brodsky, E. E., & Agnew, D. C. (2006). Seismic waves increase permeability. *Nature, 441,* 1135–1138.

Freund, L. B., & Barnett, D. M. (1976). A two dimensional analysis of surface deformation due to dip-slip faulting. *Bulletin of the Seismological Society of America, 66,* 667–675.

Fuentes Arreazola, M. A., Ramírez Hernández, J., & Vázquez González, R. (2018). Hydrogeological properties estimation from groundwater level natural fluctuations analysis as a low-cost tool for the Mexicali Valley aquifer. *Water, 10*(5), 586. https://doi.org/10.3390/w10050586.

Fuentes Arreazola, M. A., & Vázquez González, R. (2016a). Análisis temporal y frecuencial del registro de nivel de agua en el pozo G-1-17 de monitoreo del acuífero superficial en inmediaciones del Campo Geotérmico de Cerro Prieto. *Geotermia, Revista Mexicana de Geoenergía, 29*(1), 15–27.

Fuentes Arreazola, M. A., & Vázquez González, R. (2016b). Estimación de algunas propiedades geohidrológicas en un conjunto de pozos de monitoreo en el Valle de Mexicali, B.C. México. *Ingeniería del Agua, 20*(2), 87–101.

Glowacka, E. (1996). Vertical Creepment in the Southern Part of Imperial Fault (Mexico). In: *Proc. of AGU Fall Meeting*, San Francisco, p. F516.

Glowacka, E., Márquez Ramírez, V. H., Nava, F. A., Sarychikhina, O., Farfán, F., García Arthur, M. A., & Reyes, V. (2014). Slip events triggered on the Saltillo fault, Baja California, Mexico. In: *Proc. Workshop "EARTHQUAKES: nucleation, triggering, and relationships with aseismic processes"*, Cargèse, Corsica, France.

Glowacka, E., & Nava, F. A. (1996). Major earthquake in Mexicali valley, Mexico, and fluid extraction at cerro prieto geothermal field. *Bulletin of the Seismological Society of America, 86,* 93–105.

Glowacka, E., Nava Pichardo, F. A., de Cossío, Díaz, Batani, G. E., Wong Ortega, V. M., & Farfán Sánchez, F. J. (2002). Fault slip, seismicity, and deformation in Mexicali Valley, Baja California, Mexico, after the M 7.1 1999 Hector Mine earthquake. *Bulletin of the Seismological Society of America, 92*(4), 1290–1299.

Glowacka, E., Navarro León, R., Márquez Ramírez, V. H., Sarychikhina, O., Farfán, F., García Arthur, M. A., & Gálvez, O. (2017). Aplicación de los inclinómetros para medir deformaciones locales causadas por subsidencia, eventos de creep o disparadas por los sismos. In: *Proc. Reunión Anual de la Unión Geofísica Mexicana (RAUGM) 2017*, Puerto Vallarta, Mexico.

Glowacka, E., Sarychikhina, O. Suárez-Vidal, F., Nava, F. A., Farfan, F., Cossio Battani, G. D., & Guzmán, M. (2007). Aseismic slip observed on the faults in Mexicali Valley, Baja California, Mexico. In: *Proc. of AGU, Spring Meeting*.

Glowacka, E., Sarychikhina, O., Márquez Ramírez, V. H., Nava, A., Farfán, F., & García Arthur, M. A. (2015), Deformation Around the Cerro Prieto Geothermal Field Recorded by the Geotechnical Instruments Network REDECVAM during 1996–2009. In: *Proc. of World Geothermal Congress 2015*, Melbourne, Australia.

Glowacka, E., Sarychikhina, O., Suárez, F., Nava, F. A., & Mellors, R. (2010). Anthropogenic subsidence in the Mexicali Valley, Baja California, Mexico, and slip on the Saltillo fault. *Environmental Earth Sciences, 59*(7), 1515–1524.

Glowacka, E., Sarychikhina, O., Vázquez González, R., Nava, F. A., Munguía, L., Farfan, F., Diaz De Cossio, G., & Garcia Arthur, M. A. (2008). Slow aseismic slip in the pull-apart center of Cerro Prieto (Baja California, Mexico), from geotechnical instruments and InSAR observations. In: *Proc. of SSA 2008 Annual Meeting, Seismological Society of America*, Santa Fe, New Mexico, USA. *Seismological Research Letters* 79, 282.

Glowacka, E., Sarychikhina, O., Vázquez, R., Vidal, A., Aguado Guzmán, C., López Hernández, M., Farfán, F., & Díaz de Cossío Batani, G. (2011). Coseismic and Aseismic Anomalies Recorded by Geotechnical Instruments in the Cerro Prieto Pull Apart Basin, Baja California, México. In: *Proc. of SSA 2011 Annual Meeting, Seismological Society of America*, Memphis, Tennessee, USA. *Seismological Research Letters* 82(2), 353–354.

González, J., Glowacka, E., Suárez, F., Quiñones, J. G., Guzmán, M., Castro, J. M., et al. (1998). Movimiento reciente de la Falla Imperial, Mexicali, B. C. Ciencia para todos Divulgare. *Universidad Autónoma de Baja California, 6*(22), 4–15.

Grecksch, G., Roth, F., & Kumpel, H. J. (1999). Coseismic well-level changes due to the 1992 Roermond earthquake compared to static deformation of half-space solutions. *Geophysical Journal International, 138*(2), 470–478.

Igarashi, G., Saeki, S., Takahata, N., Sumikawa, K., Tasaka, S., Sasaki, Y., et al. (1995). Groundwater radon anomaly before the Kobe earthquake in Japan. *Science, 269*, 60–61.

King, Ch.-Y. (2018). Characteristics of a sensitive well showing pre-earthquake water-level changes. *Pure and Applied Geophysics*. https://doi.org/10.1007/s00024-018-1855-4.

Kitagawa, Y., & Koizumi, N. (2000). A study on the mechanism of coseismic groundwater changes: interpretation by a groundwater model composed of multiple aquifers with different strain responses. *Journal of Geophysical Research, 105*, 19121–19134.

Kitagawa, Y., Koizumi, N., & Tsukuda, T. (1996). Comparison of postseismic groundwater temperature changes with earthquake-induced volumetric strain release: Yudani hot spring, Japan. *Geophysical research letters, 23*(22), 3147–3150.

Koizumi, N., Kitagawa, Y., Matsumoto, N., Takahashi, M., Sato, T., Kamigaichi, O., et al. (2004). Preseismic groundwater level changes induced by crustal deformations related to earthquake swarms off the east coast of Izu Peninsula, Japan. *Geophysical Research Letters, 31*, L10606. https://doi.org/10.1029/2004GL019557.

Kümpel, H. J. (1991). Poroelasticity: parameters reviewed. *Geophysical Journal International, 105*, 783–799. https://doi.org/10.1111/j.1365-246X.1991.tb00813.x.

Lippmann, M. J., Goldstein, N. E., Halfman, S. E., & Witherspoon, P. A. (1984). Exploration and development of the Cerro Prieto Geothermal Field. *Journal Of Petroleum Technology, 36*(9), 1579–1591.

Lira, H. (2006). *Características del sismo del 23 de Mayo de 2006. Informe RE-023/2006.* Comisión Federal de Electricidad, Residencia de Estudios, México.

Liu, C.-Y., Chia, Y., Chuang, P.-Y., Chiu, Y.-C., & Tseng, T.-L. (2018). Impacts of hydrogeological characteristics on groundwater-level changes induced by earthquakes. *Hydrogeology Journal, 26*, 451–465. https://doi.org/10.1007/s10040-017-1684-z.

Lohman, R. B., & McGuire, J. J. (2007). Earthquake swarms driven by aseismic creep in the Salton Trough, California. *Journal of Geophysical Research, 112*, B04405. https://doi.org/10.1029/2006JB004596.

Luo, J., Sun, L., Zhang, W., Li, M., & Guo, X. (2011). Co-seismic groundwater-level and temperature changes of the 2011 Mw9. 0 Japan earthquake in Chinese mainland. *Geodesy and Geodynamics, 2*(4), 40–45.

Magistrale, H. (2002). The relation of southern San Jacinto fault zone to the Imperial and Cerro Prieto faults. *Geological Society of America Special Paper, 365*, 271–278.

Manga, M., Beresnev, I., Brodsky, E. E., Elkhoury, J. E., Elsworth, D., Ingebritsen, S. E., et al. (2012). Changes in permeability caused by transient stresses: field observations, experiments, and mechanisms. *Reviews of Geophysics, 50*, RG2004. https://doi.org/10.1029/2011rg00038.

Matsumoto, N., & Roeloffs, E. (2003). Hydrological response to earthquakes in the Haibara well, central Japan—II. Possible mechanism inferred from time-varying hydraulic properties. *Geophysical Journal International, 155*, 899–913. https://doi.org/10.1111/j.1365-246X.2003.02104.x.

McHugh, S., & Johnston, M. J. S. (1977). An analysis of coseismic tilt changes from an array in Central California. *Journal of Geophysical Research, 82*, 5692–5698.

Montgomery, D. R., & Manga, M. (2003). Streamflow and water well responses to earthquakes. *Science, 300*, 2047–2049.

Muir-Wood, R., & King, G. C. P. (1993). Hydrological signatures of earthquake strain. *Journal of Geophysical Research, 98*, 22035–22068.

Munguía, L., Glowacka, E., Suárez-Vidal, F., Lira-Herrera, H., & Sarychikhina, O. (2009). Near-fault strong ground motions recorded during the morelia normal-fault earthquakes of may 2006 in Mexicali Valley, BC, Mexico. *Bulletin of the Seismological Society of America, 99*(3), 1538–1551.

Nava, F. A., & Glowacka, E. (1999). Fault-slip triggering, healing, and viscoelastic afterworking in sediments in the Mexicali-Imperial Valley. *Pure and Applied Geophysics, 156*, 615–629.

Orihara, Y., Kamogawa, M., & Nagao, T. (2014). Preseismic Changes of the Level and Temperature of Confined Groundwater related to the 2011 Tohoku Earthquake. *Scientific Reports, 4*, 6907. https://doi.org/10.1038/srep06907.

Quilty, E. G., & Roeloffs, E. A. (1997). Water-level changes in response to the 20 December 1994 earthquake near Parkfield, California. *Bulletin of the Seismological Society of America, 87*(2), 310–317.

Rice, J., & Cleary, M. P. (1976). Some basic stress diffusion solutions for fluid-saturated elastic porous media with compressible constituents. *Reviews of Geophysics and Space Physics, 14*(2), 227–241.

Roeloffs, E. A. (1988). Hydrologic precursors to earthquakes: a review. *Pure and Applied Geophysics, 126*(2), 177–209.

Roeloffs, E. A. (1996). Poroelastic techniques in the study of earthquake-related hydrologic phenomena. *Advances in Geophysics, 37*, 135–195.

Roeloffs, E. A. (1998). Persistent water level changes in a well near Parkfield, California, due to local and distant earthquakes. *Journal of Geophysical Research, 103*, 869–889.

Sarychikhina, O., Glowacka, E., Mellors, R., & Suárez-Vidal, F. (2011). Land subsidence in the Cerro Prieto Geothermal Field, Baja California, Mexico, from 1994 to 2005. An integrated analysis of DInSAR, leveling and geological data. *Journal of Volcanology and Geothermal Research, 204*, 76–90.

Sarychikhina, O., Glowacka, E., Mellors, R., Vazquez, R., Munguia, L., & Guzman, M. (2009). Surface Displacement and Groundwater Level Changes Associated with the 24 May 2006 Mw 5.4 Morelia Fault Earthquake, Mexicali Valley, Baja California, Mexico. *Bulletin of the Seismological Society of America, 99*(4), 2180–2189.

Sarychikhina, O., Glowacka, E., & Robles, B. (2018). Multi-sensor DInSAR applied to the spatiotemporal evolution analysis of ground surface deformation in Cerro Prieto basin, Baja California, Mexico, for the 1993–2014 period. *Natural Hazards, 92*, 225–255.

Sarychikhina, O., Glowacka, E., Robles, B., Nava, F. A., & Guzman, M. (2015). Estimation of seismic and aseismic deformation in Mexicali Valley, Baja California, Mexico, in the 2006–2009 Period, using precise leveling, DInSAR, geotechnical instruments data, and modeling. *Pure and Applied Geophysics, 172*, 3139–3162.

Scholz, C. H., Sykes, L. R., & Aggarwal, Y. P. (1973). Earthquake prediction: a physical basis. *Science, 181*(4102), 803–810.

Sil, S. (2006). Response of Alaskan wells to near and distant large earthquakes. M. Sc. Thesis, University of Alaska, Fairbanks, Alaska

Sil, S., & Freymuller, J. T. (2006). Well water level changes in Fairbanks, Alaska, due to the great Sumatra-Andaman earthquake. *Earth Planets Space, 58,* 181–184.

Simpson, D. W., Leith, W. S., & Scholz, C. H. (1988). Two types of reservoir-induced seismicity. *Bulletin of the Seismological Society of America, 78,* 2025–2040.

Suárez-Vidal, F., Mendoza-Borunda, R., Nafarrete-Zamarripa, L., Rámirez, J., & Glowacka, E. (2008). Shape and dimensions of the Cerro Prieto pull-apart basin, Mexicali California, México, based on the regional seismic record and surface structures. *International Geology Review, 50*(7), 636–649.

Suárez-Vidal, F., Munguia-Orozco, L., Gonzalez-Escobar, M., Gonzalez-Garcia, J., & Glowacka, E. (2007). Surface rupture of the morelia fault near the cerro prieto geothermal Field, Mexicali, Baja California, Mexico, during the Mw 5.4 earthquake of 24 May 2006. *Seismological Research Letters, 78*(3), 394–399.

Takemoto, S. (1995). Recent results obtained from continuous monitoring of crustal deformation. *Journal of Physics of the Earth, 43,* 407–420.

Talwani, P., & Acree, C. (1985). Pore pressure diffusion and the mechanism of reservoir-induced seismicity. *Pure and Applied Geophysics, 122,* 947–965.

Tsunogai, U., & Wakita, H. (1995). Precursory chemical changes in ground water: Kobe earthquake, Japan. *Science, 269,* 61–63.

Vázquez González, R., Glowacka, E., & Díaz Fernández, A. (2005). Monitoreo continuo del nivel piezométrico en la zona geotérmica de Cerro Prieto. In: *Proc. of IV Reunión Nacional de Ciencias de la Tierra*, Juriquilla, Querétaro, Mexico.

Vázquez González, R., Ramírez Hernández, J., Martín Barajas, A., Carreón Diazconti, C., García Cueto, O. R., Miranda Reyes, F., Vázquez Hernández, F., Benítez Pérez, H., & Espinoza García, S.

(1998). Estudio Geohidrológico del Campo Geotérmico de Cerro Prieto, Mexicali, B.C. CICESE-CFE. Contrato No. RCGP-CLS-002/97.

Wakita, H. (1975). Water wells as possible indicators of tectonic strain. *Science, 189,* 553–555.

Wakita, H., Igarashi, G., & Notsu, K. (1991). An anomalous radon decrease in groundwater prior to an M6. 0 earthquake: a possible precursor? *Geophysical Research Letters, 18*(4), 629–632.

Wang, H. F. (1993). Quasi-static poroelastic parameters in rock and their geophysical applications. *Pure and Applied Geophysics, 141*(2), 269–286.

Wang, C.-Y., & Barbourb, A. J. (2017). Influence of pore pressure change on coseismic volumetric strain. *Earth and Planetary Science Letters, 275,* 152–159.

Wang, C.-Y., & Chia, Y. (2008). Mechanism of water level changes during earthquakes: near field versus intermediate field. *Geophysical Research Letters, 35,* L12402. https://doi.org/10.1029/2008GL034227.

Wang, R., Lorenzo-Martín, F., & Roth, F. (2003). Computation of deformation induced by earthquakes in a multi-layered elastic crust—FORTRAN programs EDGRN/EDCMP. *Computers & Geosciences, 29*(2), 195–207.

Wang, C.-Y., & Manga, M. (2010). Hydrologic responses to earthquakes—a general metric. *Geofluids, 10,* 206–216.

Wang, C.-Y., Manga, M., Wang, C.-H., & Chin, C.-H. (2012). Earthquakes and subsurface temperature changes near an active mountain front. *Geology, 40,* 119–122. https://doi.org/10.1130/G32565.1.

Wei, S., Avouac, J. P., Hudnut, K. W., Donnellan, A., Parker, J. W., Graves, R., et al. (2015). The 2012 Brawley swarm triggered by injection-induced aseismic slip. *Earth and Planetary Science Letters, 422,* 115–125.

Wyatt, F. K. (1988). Measurements of coseismic deformation in southern California: 1972–1982. *Journal of Geophysical Research, 93,* 7923–7942.

(Received  December 18, 2017, revised  June 8, 2018, accepted  June 11, 2018)

Pure Appl. Geophys.
© 2018 Springer International Publishing AG, part of Springer Nature
https://doi.org/10.1007/s00024-018-1800-6

# Background Stress State Before the 2008 Wenchuan Earthquake and the Dynamics of the Longmen Shan Thrust Belt

KAIYING WANG,[1] YU. L. REBETSKY,[2] XIANGDONG FENG,[3] and SHENGLI MA[1]

*Abstract*—A stress reconstruction was performed based on focal mechanisms around the Longmen Shan region prior to the 2008 $M_s$ 8.0 Wenchuan earthquake using a newly developed algorithm (known as MCA). The method determines the stress tensor, including principal axes orientations, and quantitative stress values, such as the effective confining pressure and maximum shear stress. The results of the MCA application using data recorded by the regional network from 1989 to April 2008 show the background stress state around the Longmen Shan belt before the Wenchuan earthquake. The characteristics of the stress orientation reveal that the Longmen Shan region is primarily under the eastward extrusion of the eastern Tibetan plateau. Non-uniform quantitative stress distributions show low stress levels in the upper crust of the middle Longmen Shan segment, which is consistent with the observed high-angle reverse faulting associated with the 2008 Wenchuan earthquake. In contrast, other study areas, such as the Bayankela block and the NW strip extending to the Sichuan basin, show high stress intensity. This feature coincides with heterogeneity in the wave speed image of the upper crust in this region, which shows high S-wave speed in the high stress areas and comparatively low S-wave speed in low stress areas. Deformation features across the Longmen Shan belt with the slow rates of convergence determined by GPS and the distribution of surface deformation rates also are in keeping with our stress results. We propose a dynamic model in which sloping uplift under the Longmen Shan, which partly counteracts the pushing force from the eastern plateau, causes the low-quantitative stresses in the upper crust beneath the Longmen Shan. The decreasing gravitational potential energy beneath the Longmen Shan leads to earthquake thrust faulting and plays an important role in the geodynamics of the area that results from ductile thickening of the deep crust behind the Sichuan basin, creating a narrow, steep margin.

**Key words:** Wenchuan earthquake, deformation features, MCA, quantitative stress, background stress state, low stress level, sloping uplift.

[1] State Key Laboratory of Earthquake Dynamics, Institute of Geology, China Earthquake Administration, P.O. Box 9803, Beijing 100029, China. E-mail: wangky@ies.ac.cn

[2] Institute of Physics of the Earth, Russian Academy of Sciences, Moskva 123810, Russia.

[3] Earthquake Administration of Hebei Province, Shijiazhuang 050012, China.

## 1. Introduction

The devastating May 12, 2008 Wenchuan earthquake ($M_w$ 7.9) occurred beneath the eastern margin of the Tibetan plateau, rupturing the central high-angle Longmen Shan thrust fault (Fig. 1). This observation raises an important question: what geodynamics are responsible for this special type of strong earthquake? The answer to the question is closely related to recent tectonic deformation of the eastern Tibetan plateau, which is characterized by eastern expansion as a result of the continent–continent collision of India with Eurasia. For several years, there has been controversy regarding the style of the lateral expansion of the eastern Tibetan Plateau. The argument is that the plateau may deform either by movement of rigid crustal blocks along large strike-slip faults (Tapponnier et al. 1982; Tapponnier et al. 2001), by continuous deformation (Houseman and England 1993; Copley 2008), or by the eastward flow of viscous lower crust (Royden et al. 1997, 2008).

There should be different geodynamics and measured stress states in the Longmen Shan tectonic zone corresponding to these different deformational models. The first two include extrusion along the left-lateral strike-slip faults that slice Tibet's east side or a continuous deformation style in which vertical planes deform by pure shear, and there are no vertical gradients of horizontal velocity. From these two styles, it can be inferred that the Longmen Shan tectonic zone is primarily subject to a horizontal pushing force, that the stress orientation should be consistent from the shallow to deep crust, and that there should be little change in stress intensities when lithostatic pressure is taken off. In contrast, in the lower crustal flow model, deep crustal material moves eastward from middle Tibet around the

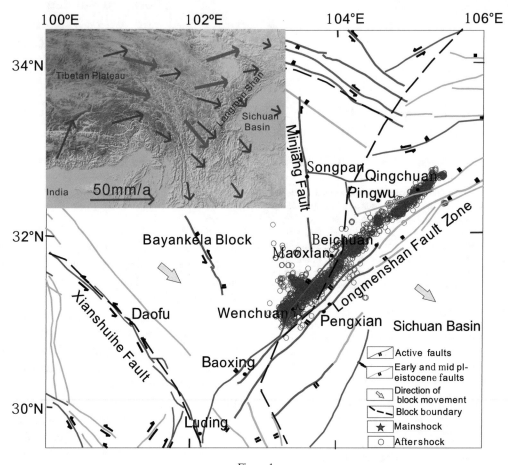

Figure 1
Sketch of the active tectonics in the Longmen Shan region including epicenters of the 2008 $M_w$ 7.9 Wenchuan earthquake sequence. In the top left corner blue arrows indicates GPS velocity vectors and gray arrows shows crustal flow

strong crust of the Sichuan basin and accumulates behind the basin. In this model, there is an obvious difference between the stress state in the upper crust and the deep crust under the Longmen Shan, and the uplift of the material piling up plays an important role in the geodynamics of the area.

The Longmen Shan has numerous geological features atypical of active convergent mountain belts (Burchfiel et al. 2008; Kirby et al. 2008), including that young high mountains reach more than 4000 m relief without adjacent foreland subsidence. Analysis of GPS velocity data (Wang et al. 2015; Zhang et al. 2008) shows that the shortening rate across the Longmen Shan is less than 3 mm/year, which is an order of magnitude lower than the crustal strain rate in the adjacent Xianshuihe fault. Rapid uplift and

slow horizontal pressing indicate that the horizontal driving force in the upper Longmen Shan crust is weak. Coseismic deformation associated with the 2008 Wenchuan earthquake occurred mostly within the Longmen Shan fault zone and decreased very rapidly away from the surface rupture zone (Xu et al. 2009; Liu-Zeng et al. 2009). Aftershocks have been confined to a particularly narrow belt along the Longmen Shan fault zone. These characteristics suggest that the horizontal force from the eastern Tibet acting on the Longmen Shan is not the dominant cause of the high-angle reverse faulting and tectonic deformation in the upper crust.

Clark and Royden (2000) proposed that crustal thickening in eastern Tibet occurs largely within a weak (low-viscosity) zone in the mid- to lower crust.

For an assumed 15-km-thick channel, Newtonian fluid model results indicate that crustal flows create the broad, gentle margins of eastern plateau and accumulate behind the Sichuan basin, creating a narrow, steep margin. Zones of weakness in the deep crust that thicken eastwards towards the craton beneath the Sichuan basin have been identified from structural imaging and interpreted as crustal flow channels (Liu et al. 2014). This interpretation can be tested by determining whether that the upper Longmen Shan crust is in a low horizontal stress, implying that the eastward pushing force of the Tibetan plateau in the upper crust acting on the Longmen Shan is small and dose not completely account for the Wenchuan earthquake.

To characterize the geodynamics of the Longmen Shan belt, it is important to know the stress state at seismogenic depths in the upper crust, where we propose a low stress level. In situ stress measurements using a hydro-fracturing technique were obtained in the northeastern Longmen Shan belt after the 2008 earthquake, indicating that the stress values at several depths within the Longmen Shan faults are considerably lower than those at Pingwu, which is located at hanging wall northwest of the Longmen Shan (Chen et al. 2012). However, because of the small number of measurement points and the limited depth of 400 meters, the stress intensity of the Longmen Shan belt at seismogenic depth remains poorly understood.

The abundance of small earthquakes offers many different focal mechanisms, which can be used to obtain principal stress axes orientations in different sub-regions under the assumption of uniform stress. The most commonly used approaches for obtaining principal stress axes orientations are described by Angelier (1979, 1989), Ellsworth and Xu (1980), Gephart and Forsyth (1984), Michael (1987), and Xu et al. (1992). Other methods for estimating deviatoric stress magnitudes using focal mechanisms before and after an earthquake and coseismic stress change have been proposed (Wan et al. 2006; Hardebeck 2012; Yang et al. 2013).

## 2. Method

As a continuation of the work of Angelier and others, a method of cataclastic analysis called MCA,

involving two basic steps, was developed to determine complete stress tensor components based on earthquake focal mechanisms (Rebetsky et al. 2012, 2016). The orientation of the principal stress axes and the Lode–Nadai coefficient or stress ratio, which characterizes the shape of the stress ellipsoid, are evaluated during the first stage of the MCA. The first stage also involves identifying a homogeneous sample set of earthquake focal mechanisms that characterizes the quasi-homogeneous deformation of a specific crustal domain.

The procedures in the first stage are very similar to those in the classical method developed by Angelier (1989, 1990), Carey-Gailhardis and Mercier (1987), and Gushchenko and Kuznetsov (1979). The procedures for generating homogeneous samples of earthquake focal mechanisms are based on inequalities that are similar to the one used in the right dihedral method by Angelier and Mechler (1977). In this study, the homogeneous sample includes events where the quadrants formed by the nodal planes intersect. This is equivalent to requiring that the irreversible plastic deformations of elongation and shortening along a principal axis due to each earthquake ($\alpha$) are defined according to the axis' index. Elongation accumulates only along the $\sigma_1$ axis (minimum compression), and shortening accumulates only along the $\sigma_3$ axis (maximum compression). Therefore,

$$d\varepsilon_{11}^\alpha \geq 0, \quad d\varepsilon_{33}^\alpha \geq 0. \tag{1}$$

$\sigma_1$ and $\sigma_3$ correspond to the algebraically largest and smallest of the principal stresses, respectively.

Instead of the inequalities in Eq. (1), the MCA has

$$d\varepsilon_{11}^\alpha \geq d\varepsilon_{22}^\alpha \geq d\varepsilon_{33}^\alpha \tag{2}$$

$$d\varepsilon_{11}^\alpha + d\varepsilon_{22}^\alpha + d\varepsilon_{33}^\alpha = 0, \tag{3}$$

where $d\varepsilon_{ii}^\alpha$ are components of increment of plastic deformations of elongation and shortening in directions of principal stress axis $\sigma_i$ ($i = 1, 2, 3$) after earthquake number $\alpha$ from a homogeneous sample.

Condition (2) is a consequence of the ordering principle of the development of irreversible deformations used in the theory of plasticity. This principle follows from the postulate of maximum plastic dissipation (Mises 1928). Correct stresses are those that maximize the elastic energy dissipation for a known irreversible deformation tensor.

In the MCA, inequality (2) is assumed as the criterion for compiling homogeneous sets of earthquake mechanisms that are used to define the parameters of the stress tensor for quasi-homogeneous domains. After the right dihedral method (Angelier and Mechler 1977), using expression (2) and data on earthquake mechanisms, it is possible to locate the principal stress axes of the desired stress tensor on the unit hemisphere areas.

After the first stage of the MCA, out of six components of stress tensor, the confining pressure p and the module of maximal shear stress $\tau$ are left undefined. The approach to determining the effective isotropic pressure $p^*$ (tectonic pressure without fluid pressure) and maximum shear stress $\tau$ is based on experiments on the brittle fracture of rock samples (Byerlee 1968, 1978), according to which a brittle fracture band is allocated on the Mohr diagram (Rebetsky et al. 2012). All points on Mohr's diagram characterizing the critical state on newly developed and re-activated cracks fall into the area between upper yield envelope (inner brittle strength) and the bottom line of resistance to static friction with minimal—zero cohesion (Fig. 2).

At second stage of the MCA, after calculating the reduced stresses from a homogeneous sample, and applying the corresponding points to the Mohr diagram, we can estimate the effective pressure $p^{\text{eff}}$ and maximum shear stress $\tau$ normalized to the unknown cohesive strength ($T_f$) of the rock massif (Rebetsky et al. 2012, 2016).

$$\frac{p^{\text{eff}}}{\tau_f} = \frac{\left(\tilde{\sigma}_{nt}^{K} - k_s \tilde{\sigma}_{nn}^{K} - k_s \tilde{\mu}_\sigma/3\right)}{k_s\left[\sec(2\beta_s) - \left(\tilde{\tau}_{ns}^{K} - k_s \tilde{\sigma}_{nn}^{K}\right)\right]};$$

$$\frac{\tau}{\tau_f} = \frac{1}{\sec(2\varphi_s) - \left(\tilde{\sigma}_{ns}^{K} - k_s \tilde{\sigma}_{nn}^{K}\right)}. \tag{4}$$

$\tilde{\sigma}_{nn}$ and $\tilde{\tau}_n$ are the corresponding reduced shear stress and effective normal stress acting on the crack plane from homogeneous sample of slip fault sets:

$$\tilde{\sigma}_{nn} = \left(\sigma_{nn} + p^{\text{eff}}\right)/\tau = (1 - \mu_\sigma)\left(n_1^l\right)^2 - (1 + \mu_\sigma)\left(n_3^l\right)^2 + 2\mu_\sigma/3;$$

$$\tilde{\tau}_n = \tilde{\sigma}_{nt} = \sigma_{nt}/\tau = (1 - \mu_\sigma)n_1^l t_1^l - (1 + \mu_\sigma)n_3^l t_3^l; \tag{5}$$

where $\tau = (\sigma_1 - \sigma_3)/2$, $p = -(\sigma_1 + \sigma_2 + \sigma_3)/3$, $p^{\text{eff}} = p - p_{\text{fl}}$, $\beta_s = \frac{1}{2}\arctan\frac{1}{k_s}$, $\varphi_s = \arctan k_s$, $k_s$

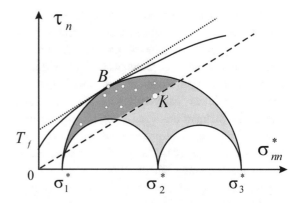

Figure 2

Zone of brittle destruction on Mohr diagram. The simplified form of a zone of the brittle destruction, are used in MCA algorithm. Solid line is Mohr failure envelope for real geo-material. Upper dotted direct line corresponds to the approximation of the Mohr failure envelope curve. Lower long-dashed line corresponds to the minimum of static friction stresses. Area of light gray color inside the big and above small circles of Mohr, defines stress states on the different orientation planes. Area of dark gray color means failure zone for restriction. Points within dark gray area are normal and shear stresses on the planes of slip faults from a homogeneous sample. The point K lies on the line of the minimum resistance of friction ($i = 1, 2, 3$) and is effective principal and normal stresses accordingly

represents the static frictional coefficient of the faults, $\mu_\sigma$ is the Lode–Nadai coefficient characterizing shape of the stress tensor, $n_i$ and $t_i$ ($i = 1, 2, 3$) are the direction cosines of the displacement vector and the shear stress on the fault plane in a coordinate system referenced to the principal axis of the stress tensor. An approach suggested in Angelier's paper (1989) should be considered to be closest to the MCA algorithm.

## 3. Data and Application of the MCA

The ratios of S to P amplitudes, as recorded on vertical component seismographs near an earthquake provide a means of determining the focal mechanism at short epicentral distances (Kisslinger 1980; Kisslinger et al. 1981). Systematic variations in S/P amplitude ratios are expected because P-wave amplitudes are large near the P and T axes of the focal mechanism and smaller near the nodal planes, whereas the S-wave amplitudes are largest near the nodal planes. Hardebeck and Shearer (2003) tested that the observed S/P ratios are generally consistent

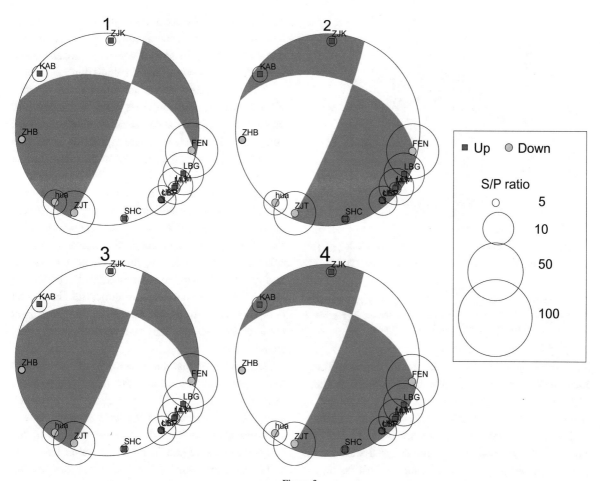

Figure 3
Example showing how to use SV/P amplitudes ratios in addition to P-wave polarities to determine an earthquake focal mechanism

with the expected mechanisms. Others (Snoke et al. 1984; Rau et al. 1996; Shen et al. 1997) developed programs to determine focal mechanism solutions by incorporating P-wave polarities and SV/P amplitude ratios. The use of the amplitude ratio along with polarity can constrain the mechanism solutions more effectively than the use of polarity alone.

Liang and Li (1984) used the SV/P amplitude ratios in addition to P-wave polarities to determine the focal mechanism of small earthquakes. The method calculates theoretical maximum SV/P amplitudes ratios from synthetic seismograms for a point source of dislocation in a planar-layered medium and obtains the best focal parameters by fitting the observed amplitude ratios with the theoretical ones. If the total number of consistent amplitude ratios is within a preset error allowance, then the

solution is accepted. Lin et al. (1991) conducted artificial data tests to prove the reliability of the method, even if the stations are not well distributed. Figure 3 is an example showing how to obtain the desired focal mechanism using the method. The serial numbers 1–4 are the best possible solutions that satisfy a predetermined data consistency criterion. The nodal planes are shown in the beach balls, and the observations are plotted as circles. When P-wave polarities are included, then solution No. 2 or No. 4 is declared to be valid in the example.

Using the method based on SV/P amplitudes combined with P-wave first-motion polarities, focal mechanism solutions were determined for 675 events recorded by at least 6 stations (Fig. 4). These events were from 1989, when the temporal distribution of seismic events was monthly in Sichuan province, to

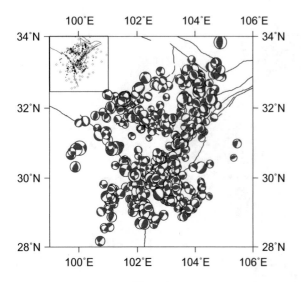

Figure 4
Focal mechanism solutions of small earthquakes recorded between 1994 and April 2008 in the Longmen Shan region. The green diamonds represent locations of station

April 2008, prior to the Wenchuan earthquake. The magnitudes for these focal mechanisms were on the scale of $M_s = 3$–5 and focal depths were distributed from 1 to 25 km. Regional digital seismic stations within Sichuan province are distributed as shown in Fig. 4. The data obtained on the stressed state correspond to the long-period component, that is, they are the average stresses for the whole period of time for the catalog of mechanisms of earthquake foci. Time variations in stresses in this work have not been investigated.

Based on the mechanism solutions, background stress tensors around the Longmenshan belt were reconstructed prior to the Wenchuan earthquake using the MCA. According to the spatial distribution of focal mechanisms and the depths of approximately 1–25 km, the grid size was chosen to be $0.2° \times 0.2°$ located at a 13 km depth, above the 2008 Wenchuan focal depth ($\sim$ 18 km) (Chen et al. 2009). The static frictional coefficient is determined to be 0.6, and rock cohesion to be 0.1 Kbar. Each homogeneous sub-region includes at least six earthquakes, and the maximum radius around a node does not exceed 35 km, which represents the elastic unloading range calculated according to earthquake magnitude (Rebetsky et al. 2012). The maximum radius should not be set too large, even if more stress points will be

obtained, otherwise some earthquakes may influence too large an area. The maximum radius should be chosen according to spatial density of focal mechanism data after several initial calculations.

The results (Fig. 5) include orientations of principal stress axes, stress regimes types, and normalized values of maximum shear stress $\tau$ relative to rock cohesion, which can represent the stress level of a unit body on a Mohr diagram. From the orientation of the principal stress axes, it is possible to regionalize the crust based on its geodynamic or stress state regimes. These regimes are defined by the shared orientation of principal stress axes and zenith direction. The types of geodynamic regimes represent the spherical octant including the horizontal extension regime, the horizontal extension with horizontal shear, the horizontal shear regime, the horizontal compression with horizontal shear, the horizontal compression regime, and the vertical shear stress state.

Directions of maximum compressive stresses in the study area are in general agreement with previous estimates (Zoback 1992; Xu et al. 1992; Heidbach et al. 2010). Distribution of the stress regimes, as well as stress axes, indicate that the upper crust of the Longmen Shan region is primarily subject to eastward horizontal compressive forces from the pushing of the upper crust away from the central Tibetan plateau and toward the eastern plateau region.

Normalized maximum shear stresses show that stress levels are very heterogeneous and, in certain areas, are comparatively large. The NW-oriented strip, from the southern Xianshuihe fault to the intersection of the Xianshuihe fault and the Longmen Shan fault, extending southeastwards to the Sichuan basin, corresponds to the highest stress level. In the Bayankela block of eastern Tibet, the stress distribution shows that most of the stresses are very high in magnitude. In contrast to these areas, the Longmen Shan area is characterized by a low stress level, with two localized patches of high stress at both ends.

The high stress distribution in the area intersecting the Xianshuihe fault and the Longmen Shan, and extending southeastward to the Sichuan basin, corresponds to a zone of comparatively high S-wave velocity ($\sim$ 3.9 km/s) at the depth of 10 km in the upper crust (Liu et al. 2014). In addition, stresses

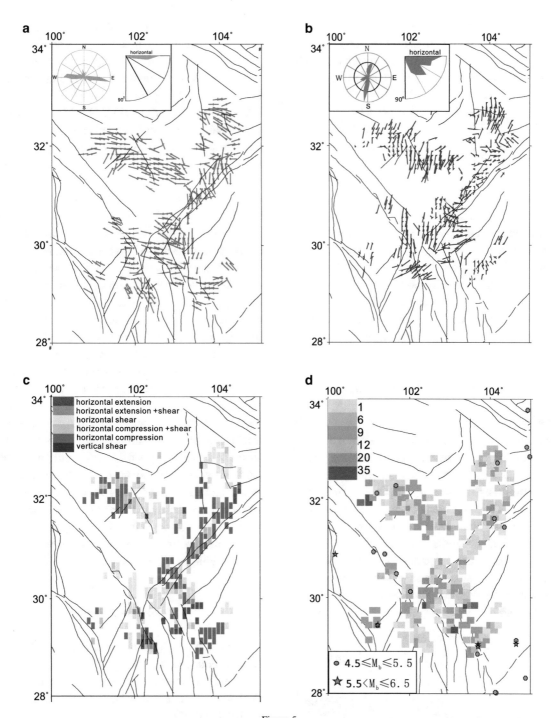

Figure 5

Stress field distribution in the Longmen Shan region: **a** projections of maximum compressive axes in the horizontal plane; **b** projections of maximum extensional axes in the horizontal plane; **c** types of stress regimes: green represents horizontal shear, blue represents horizontal compression, red represents horizontal extension; **d** normalized values of maximum shear stresses within eastern Tibet are higher than those within the Longmen Shan, similar to the wave speed heterogeneity of 3.7 vs. 3.5 km/s. The two patches of high stress in Longmen Shan also correspond to high wave speeds of approximately 3.7–3.9 km/s.

Similarly, distribution of the geodetic strain rate obtained at the decadal scale in the Sichuan and Yunnan areas shows higher strain rates in the eastern Plateau than in the Longmen Shan (Copley 2008; Wang et al. 2015), and the heterogeneity of the strain rate distribution in west Sichuan is generally consistent with the stress distribution.

The combination of variance of quantitative stress, with wave speed being in the upper crust, and the surface strain rate favors the stress inversion results. From preliminarily comparative analysis, it can be inferred that the change in rock density due to the stress and strain has an important influence on the wave speed variation in the upper crust.

## 4. Discussion and Conclusions

Orientations of stress axes and geodynamic regimes representative of the stress state at a depth of approximately 13 km beneath the Longmen Shan, as presented above, show that the upper Longmen Shan crust is primarily stressed by the pushing force from the eastern Tibetan plateau. Obviously, low stress levels along the Longmen Shan compared with other areas reveal that material accumulation under the Longmen Shan plays an important role in the geodynamics of the area, which results from ductile thickening of the deep crust behind the Sichuan basin, creating a narrow, steep margin.

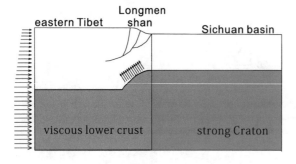

Figure 6
Conceptual cartoon to illustrate geodynamics beneath the Longmen Shan. Scales for the diagram is arbitrary

The stress state in the upper crust beneath the Longmen Shan is accommodated by a combination of plateau expansion and sloping uplift of the deeper crust beneath the Longmen Shan. The orientation of the uplift is declining towards northwest as illustrated in the conceptual cartoon (Fig. 6), which is caused by accumulation of material, similar to the decoupled pure-shear crustal thickening model proposed by Feng et al. (2016), where east-dipping ductile shear zones in the depth range $\sim$ 20 km were revealed by deep-sounding seismic reflection profiling. The uplift in part counteracts the pushing force from the eastern plateau, causing the low-quantitative stresses in the upper crust beneath the Longmen Shan, and reducing the gravitational potential energy beneath the Longmen Shan, leading to earthquake thrust faulting. Geological features such as high mountains reaching more than 4000 m relief, without adjacent foreland subsidence, and with only slow active convergence can also be interpreted from the geodynamics.

Regarding quantitative normalized stresses in the different nodes or areas concerned, it can be seen from Eq. (4) that the effective stress and the maximum shear stress are normalized to rock cohesion $\tau_f$, which may differ for different rock physics and for different scales of stress field. As a result, normalized stresses depend on the tectonic development of a region and its current state, so the acquired stress intensities for the region are only semi-quantitative, and the low stress level in upper Longmen Shan crust may be reduced to a certain extent due to cataclastic fault material or other reasons.

## Acknowledgements

This research is supported by the National Natural Science Foundation of China (Grant no. 41572181) and the Basic Research Funds from the Institute of Geology, China Earthquake Administration (Grant no. IGCEA1605).

## References

Angelier, J. (1979). Determination of mean principal directions of stresses for a given fault population. *Tectonophysics, 56,* T17–T26.

Angelier, J. (1989). From orientation to magnitude in paleostress determinations using fault slip data. *Journal of Structural Geology, 11*(1-2), 37–49.

Angelier, J. (1990). Inversion field data in fault tectonics to obtain the regional stress—III. A new rapid direct inversion method by analytical means. *Geophysical Journal International, 10,* 363–367.

Angelier, J., & Mechler, P. (1977). Sur une methode graphique de recherche des contraintes principales egalement utilisable en tectonique et en seismologie: la methode des diedres droits. *Bulletin de la Société géologique de France XIX, 7*(6), 1309–1318.

Burchfiel, B. C., Royden, L. H., van der Hilst, R. D., et al. (2008). A geological and geophysical context for the Wenchuan earthquake of 12 May 2008, Sichuan, People's Republic of China. *GSA Today, 18,* 4–11.

Byerlee, J. D. (1968). Brittle-ductile transition in rocks. *Journal of Geophysical Research, 73*(14), 4741–4750.

Byerlee, J. D. (1978). Friction of rocks. *Pure and Applied Geophysics, 116,* 615–626.

Carey-Gailhardis, E., & Mercier, J. L. (1987). A numerical method for determining the state of stress using focal mechanisms of earthquake populations: Application to Tibetan teleseismic and microseismicity of Southern Peru. *Earth and Planetary Science Letters, 82,* 165–179.

Chen, Q. C., Feng, C. J., Meng, W., et al. (2012). Analysis of in situ stress measurements at the northeastern section of the Longmenshan fault zone after the 5.12 Wenchuan earthquake. *Chinese Journal of Geophysics, 55*(12), 3923–3932.

Chen, J. H., Liu, Q. Y., Li, S. C., et al. (2009). Seismotectonic study by relocation of the Wenchuan M_s 8.0 earthquake sequence. *Chinese Journal of Geophysics, 52*(2), 390–397.

Clark, M. K., & Royden, L. H. (2000). Topographic ooze: Building the eastern margin of Tibet by lower crustal flow. *Geology, 28,* 703–706.

Copley, A. (2008). Kinematics and dynamics of the southeastern margin of the Tibetan Plateau. *Geophysical Journal International, 174,* 1081–1100.

Ellsworth, W. L., & Xu, Z. H. (1980). Determination of the stress tensor from focal mechanism data. *EOS Transactions AGU, 61,* 1117.

Feng, S. Y., Zhang, P. Z., Liu, B. J., et al. (2016). Deep crustal deformation of the Longmen Shan, eastern margin of the Tibetan Plateau, from seismic reflection and finite element modeling. *Journal of Geophysical Research: Solid Earth, 121,* 767–787.

Gephart, J. W., & Forsyth, D. W. (1984). An improved method for determining the regional stress tensor using earthquake focal mechanism data: Application to the San Fernando earthquake sequence. *Journal of Geophysical Research, 89,* 9305–9320.

Gushchenko, O. I., & Kuznetsov, V. A. (1979). Determination of the orientations and the ratio of principal stresses on the basin of tectonic fault slip data. In *Stress fields in the lithosphere*, Nauka, Moscow (pp. 60–66) **(in Russian)**.

Hardebeck, J. L. (2012). Coseismic and postseismic stress rotations due to great subduction zone earthquakes. *Geophysical Research Letters, 39,* L21313.

Hardebeck, J. L., & Shearer, M. (2003). Using S/P amplitude ratios to constrain the focal mechanisms of small earthquakes. *Bulletin of the Seismological Society of America, 93*(6), 2434–2444.

Heidbach, O., Tingay, M., Barth, A., et al. (2010). Global crustal stress pattern based on the World Stress Map database release 2008. *Tectonophysics, 482,* 3–15.

Houseman, G., & England, P. (1993). Crustal thickening versus lateral expulsion in the Indian–Asian continental collision. *Journal of Geophysical Research, 98,* 12233–12249.

Kirby, E., Whipple, K., & Harkins, N. (2008). Topography reveals seismic hazard. *Nature Geoscience, 1*(8), 485–487.

Kisslinger, C. (1980). Evaluation of S to P amplitude ratios for determining focal mechanisms from regional network observations. *Bulletin of the Seismological Society of America, 70,* 999–1014.

Kisslinger, C., Bowman, J. R., & Koch, K. (1981). Procedures for computing focal mechanisms from local (SV/P) data. *Bulletin of the Seismological Society of America, 71*(6), 1719–1729.

Liang, S. H., & Li, Y. (1984). On the determining of source parameters of small earthquakes by using amplitude ratios of P and S from regional network observations. *Chinese Journal of Geophysics, 27*(3), 249–257.

Lin, J. Z., Jiang, W. Q., Li, Y. M., et al. (1991). Determination of source parameters of small earthquakes in the east part of Guangdong and South part of Fujian province. *Acta Seismologica Sinica, 13*(4), 420–429.

Liu, Q. Y., Hilst, Robert V. D., Li, Y., et al. (2014). Eastward expansion of the Tibetan Plateau by crustal flow and strain partitioning across faults. *Nature Geoscience*. https://doi.org/10.1038/ngeo2130.

Liu-Zeng, J., Zhang, Z. Q., Wen, L., et al. (2009). Co-seismic ruptures of the 12 May 2008, Ms 8.0 Wenchuan earthquake, Sichuan: East–west crustal shortening on oblique, parallel thrusts along the eastern edge of Tibet. *Earth and Planetary Science Letters, 286,* 355–370.

Michael, A. J. (1987). Use of focal mechanisms to determine stress: A control study. *Journal of Geophysical Research, 89,* 11517–11526.

Rau, R. J., Wu, F. T., & Shin, T. C. (1996). Regional network focal mechanism determination using 3D velocity model and SH/P amplitude ratio. *Bulletin of the Seismological Society of America, 86*(5), 1270–1283.

Rebetsky, Yu L, Kuchai, O. A., Sycheva, N. A., et al. (2012). Development of inversion methods on fault slip data: Stress state in orogenes of the Central Asia. *Techtonophysics, 581,* 114–131.

Rebetsky, Yu L, Polets, A. Y., & Zlobin, T. K. (2016). The state of stress in the Earth's crust along the northwestern flank of the Pacific seismic focal zone before the Tohoku earthquake of 11 March 2011. *Tectonophysics, 685,* 60–76.

Royden, L. H., Burchfiel, B. C., King, R., et al. (1997). Surface deformation and lower crustal flow in eastern Tibet. *Science, 276,* 788–790.

Royden, L. H., Burchfiel, B. C., & van der Hilst, R. D. (2008). The geological evolution of the Tibetan Plateau. *Science, 321,* 1054–1058.

Shen, Y., Forsyth, D. W., Conder, J., & Dorman, L. M. (1997). Investigation of microearthquake activity following an intraplate teleseismic swarm on the west flank of the southern East Pacific Rise. *Journal of Geophysical Research, 102*(B1), 459–475.

Snoke, J. A., Munsey, J. W., Teague, A. G., & Bollinger, G. A. (1984). A program for focal mechanism determination by combined use of polarity and SV-P amplitude ratio data. *Earthquake Notes, 55*(3), 15.

Tapponnier, P., Peltzer, G., Dain, A. Y. L., et al. (1982). Propagating extrusion tectonics in Asia: New insights from simple experiments with plasticine. *Geology, 10,* 611–616.

Tapponnier, P., Xu, Z., Roger, F., et al. (2001). Oblique stepwise rise and growth of the Tibet Plateau. *Science, 294,* 1671–1677.

von Mises, R. (1928). Mechanik der plastischen Formänderung von Kristallen. *Zeitschrift für Angewandte Mathematik und Mechanik, 8,* 161–185.

Wan, Y. G., Shen, Z. K., & Lan, C. X. (2006). Deviatoric stress level estimation according to principle axes rotation of stress field before and after large strike-slip type earthquake and stress drop. *Chinese Journal of Geophysics, 49*(3), 838–844.

Wang, F., Wang, M., Wang, Y. Z., & Shen, Z. K. (2015). Earthquake potential of the Sichuan-Yunnan region, western China. *Journal of Asian Earth Sciences, 107,* 232–243.

Xu, Z. H., Wang, S. Y., Huang, Y. R., et al. (1992). Tectonic stress field of China from a large number of small earthquakes. *Journal of Geophysical Research, 97*(B8), 11867–11877.

Xu, X., Wen, X., & Yu, G. (2009). Coseismic reverse- and oblique-slip surface faulting generated by the 2008 Mw 7.9 Wenchuan earthquake China. *Geology, 37,* 515–518.

Yang, Y. R., Johnson, K. M., & Chuang, R. Y. (2013). Inversion for absolute deviatoric crustal stress using focal mechanisms and coseismic stress changes: The 2011 M9 Tohoku-oki, Japan, earthquake. *Journal of Geophysical Research, 118,* 5516–5529.

Zhang, P. Z., Xu, X. W., Wen, X. Z., et al. (2008). Slip rates and recurrence intervals of the Longmen Shan active fault zone, and tectonic implications for the mechanism of the May 12 Wenchuan earthquake, 2008, Sichuan, China. *Chinese Journal of Geophysics, 51*(4), 1066–1073.

Zoback, M. L. (1992). First and second order patterns of stress in the lithosphere: The World Stress Map Project. *Journal of Geophysical Research: Solid Earth, 97,* 11703–11728.

(Received December 20, 2016, revised August 12, 2017, accepted February 3, 2018)

Pure Appl. Geophys.
© 2017 Springer International Publishing AG
DOI 10.1007/s00024-017-1596-9

| Pure and Applied Geophysics

CrossMark

# Tectonically Induced Anomalies Without Large Earthquake Occurrences

ZHEMING SHI,[1,2] GUANGCAI WANG,[1,2] CHENGLONG LIU,[3] and YONGTAI CHE[3]

*Abstract*—In this study, we documented a case involving large-scale macroscopic anomalies in the Xichang area, southwestern Sichuan Province, China, from May to June of 2002, after which no major earthquake occurred. During our field survey in 2002, we found that the timing of the high-frequency occurrence of groundwater anomalies was in good agreement with those of animal anomalies. Spatially, the groundwater and animal anomalies were distributed along the Anninghe–Zemuhe fault zone. Furthermore, the groundwater level was elevated in the northwest part of the Zemuhe fault and depressed in the southeast part of the Zemuhe fault zone, with a border somewhere between Puge and Ningnan Counties. Combined with microscopic groundwater, geodetic and seismic activity data, we infer that the anomalies in the Xichang area were the result of increasing tectonic activity in the Sichuan–Yunnan block. In addition, groundwater data may be used as a good indicator of tectonic activity. This case tells us that there is no direct relationship between an earthquake and these anomalies. In most cases, the vast majority of the anomalies, including microscopic and macroscopic anomalies, are caused by tectonic activity. That is, these anomalies could occur under the effects of tectonic activity, but they do not necessarily relate to the occurrence of earthquakes.

**Key words:** Earthquake prediction, macroscopic anomaly, animal anomaly, groundwater, tectonic activity.

## 1. Introduction

Earthquake prediction is a worldwide conundrum, and the subject of whether earthquakes can be predicted has been debated for decades (Scholz et al.

**Electronic supplementary material** The online version of this article (doi:10.1007/s00024-017-1596-9) contains supplementary material, which is available to authorized users.

[1] State Key Laboratory of Biogeology and Environmental Geology and MOE Key Laboratory of Groundwater Circulation and Environmental Evolution, China University of Geosciences, Beijing, China. E-mail: szm@cugb.edu.cn
[2] School of Water Resources and Environment, China University of Geosciences, Beijing 100083, China.
[3] Institute of Geology, China Earthquake Administration, Beijing 100029, China.

1973; Lomnitz 1994; Geller 1997; Wyss 1997; Peresan et al. 2005). Some researchers remain pessimistic about the long-term potential for prediction based on the nonlinear nature of earthquake dynamics (Geller 1997; Yan 1997), the lack of observed precursors during the long-term earthquake prediction experiment at Parkfield (Bakun et al. 2005), and the lack of precursors before many large earthquakes (Ingebritsen and Manga 2014). Others are optimistic about earthquake prediction because of the following reasons: (1) laboratory rock experiments have demonstrated that rock dilatancy occurs under high deviatoric stresses and leads to eventual large-scale rupture, which could explain many precursors (Wang and Manga 2010; Wyss 1997); (2) many significant precursors have been observed over the last few decades (King et al. 2000; Woith 2015; Roeloffs 1988; Skelton et al. 2014; Toutain and Baubron 1999; Wallace and Teng 1980; Zhang and Fu 1981); (3) several earthquakes have been successfully predicted using precursors (e.g., 1974.2.4 Haicheng M7.3; 1975.5.29 Longling M7.3; 1976 Songpan-Pingwu M7.2) (Che and Liu 2008; Wang et al. 2006; Roeloffs 1988). In spite of these debates, efforts have been taken to detect earthquake precursors, and a substantial amount of hydrological anomalies have been reported leading up to earthquakes worldwide (Hartmann and Levy 2005; Woith 2015; Matsumoto and Koizumi 2013; Manga and Wang 2015; Skelton et al. 2014). All of these "potential earthquake anomaly signals" provide new insight into the physics of earthquakes and the tectonic processes that lead to earthquakes.

Precursor anomalies can generally be categorized into "microscopic anomaly precursors" and "macroscopic anomaly precursors." Macroscopic anomaly precursors are earthquake-related phenomena that can be detected by human senses or by simple instrumentation, while microscopic anomaly precursors are earthquake-related phenomena that can only be detected by precise instrumentation (Zhu and Wu 1982; Rikitake 1994). Macroscopic anomaly phenomena include changes in groundwater, such as the water level, flow rate, color, and taste, unusual sounds, fireballs in the air, and abnormal behaviors exhibited by animals, including the nervousness of cats, frogs, dogs, snakes, and fish (Gold and Soter 1984; Che and Yu 2006). Particularly, macroscopic groundwater anomalies appear in an area where a marked crustal deformation can be expected to occur (Rikitake 1994). In addition, macroscopic anomalies are considered as one of the key factors in short-term earthquake prediction. In China, macroscopic anomalies have played a key role in successfully predicting several earthquakes, especially in determining their location and time (e.g., 1974.2.4 Haicheng M7.3; 1975.5.29 Longling M7.3; 1976 Songpan-Pingwu M7.2) (Wallace and Teng 1980; Liu et al. 2004), despite the failed prediction of the 1976.7.28 M7.8 Tangshan earthquake. There also exist large-scale macroscopic anomalies, especially when considering groundwater anomalies (Zhang and Li 1977). Thus, macroscopic anomalies are considered as a significant signal of short-term precursors of impending earthquakes in the practice of predicting earthquakes (Wang and Shen 2010; Roeloffs 1988; Liu et al. 2004; Che et al. 2003). However, earthquakes do not always follow macroscopic anomalies, even if large-scale macroscopic anomaly precursors have occurred both spatially and temporally. These cases are especially important for us to understand the nature of macroscopic anomalies and earthquake processes. In this study, we report a case wherein large-scale anomalies, particularly macroscopic anomalies, occurred from May 2002 to June 2002 in the Xichang area of the southwestern Sichuan Province, China, but no expected large earthquake occur.

## 2. Geological and Tectonic Setting

The study area is located in the southwestern Sichuan Province and ranges from the city of Mianning in the north through the city of Xichang to the city of Qiaojia in the south. This area is located at the southeast margin of the Tibetan Plateau within a tectonically active block known as the Sichuan-Yunnan

block or Chuandian block (Fig. 1). The eastern margin of the block is composed of the Xianshuihe, Anninghe, Zemuhe, and Xiaojiang faults in order from the northwest to the southeast, and the Sichuan–Yunnan block represents one of the most active and seismic zones of the region. These four strike-slip faults trend NW–SE, approximately S–N, NNW–SSE and nearly S–N, respectively, and intersect at the Shimian, Xichang, and Ningnan areas. According to Wen and Cheng (1992), the study area can be divided into three sections based on the geometric characteristics, slip rates, seismic activity, and historical earthquakes along the faults. The three sections are as follows: the northern part of the Anninghe fault (the cities of Mianning and Xichang, Sect. 1), the Zemuhe Fault (the city of Puge, Sect. 2), and the Ning-Qiao pull-apart structure (the cities of Ningnan and Qiaojia, Sect. 3). With a length of 75 km, Sect. 1 has a slip rate of 5 ± 1 mm/year. Section 2 has a length of 90 km and a slip rate of 7 ± 1.5 mm/year. Finally, Sect. 3 has a length of 50 km, with a vertical slip rate of 1.5–2 mm/year. Their complex geological settings and intense tectonic activity lead to a high frequency of intense seismic activity. Five earthquakes with magnitudes larger than 6 have occurred along the fault zone since 1500, 3 of which were larger than 7 (Mu 2008). Wen and Cheng (1992) suggested a high potential for large earthquakes (M 6.7) in Sect. 1 and a high potential for medium earthquakes (M 4.7–M 5.5) in Sect. 3, whereas there is only a low possibility of strong earthquakes in Sect. 2 (Wen and Cheng 1992). A seismic gap (M > 4) around the Anninghe–Zemuhe fault zone was disrupted in 2001. A series of earthquakes occurred in the area at that time (2001.3.7 Butuo-Puge M 4.5; 2002.3.3 Mianning M 4.4; 2002.4.10 Puge M 4.6; 2002.5.23 Butuo M 4.0) (Fig. 1). This may reflect changes in the status of stress in the Anninghe–Zemuhe fault zone.

## 3. Anomaly Behaviors in the Study Area

### 3.1. Overall Introduction of the Macro- and Microscopic Anomaly Phenomena in the Xichang Area

The series of small-to-moderate seismic activity observed, since 2001 indicates a rising risk of strong

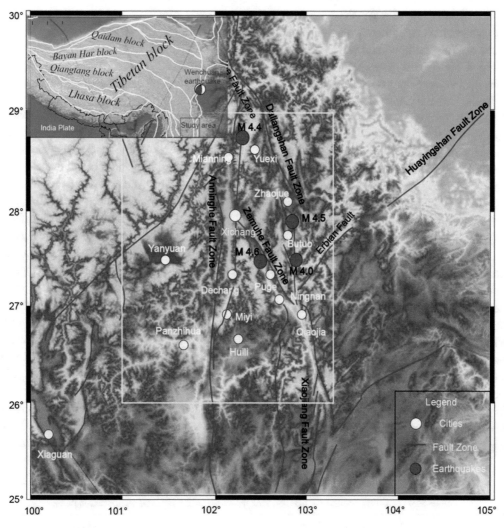

Figure 1
Geological setting of the study area

earthquakes for the future. Thus, seismologists in Sichuan Province have paid special attention to this area since that time. Beginning in March and April of 2002, a series of microscopic groundwater, geodetic, and seismic anomalies appeared in the southern Sichuan and northern Yunnan regions (Wang et al. 2006). In particular, following the 2002.4.10 Puge M 4.9 earthquake, hundreds of macroscopic anomalies were observed and reported both by local residents and amateur observers in Xichang and other towns nearby (Mu 2008). By mid-May, a tendency of increasing micro-earthquake activity was observed. According to previous experiences, a strong earthquake should occur after the foreshock sequence

(Wang et al. 2006). To identify the nature of these phenomena and their possible relationships to tectonic activity or their possible role as precursors for future earthquakes, a field investigation was conducted by a professional earthquake research group with members from the China Earthquake Administration (CEA), the Sichuan Province Earthquake Administration and the seismological center station of Xichang. The investigation lasted one month (May 25th, 2002–June 25th, 2002). The weather conditions were similar to those of the past several years, and no abnormal weather occurred during our stay. The responsibilities of this investigation include the following: analyzing the microscopic anomaly phenomena, identifying whether the macroscopic anomalies are precursors, and collecting and analyzing water samples in the important wells and springs. Some of the authors of this paper also took part in the field investigation, and our responsibilities were to mainly focus on the macroscopic groundwater anomalies. The methodologies used in the field investigation included (1) visiting local people who had witnessed what occurred and knew the background, (2) observing and recording the phenomena in the field, (3) measuring the parameters of the site (i.e., well, spring, and borehole) and groundwater (the level and temperature), and (4) sampling.

During the investigation, more than 80 macroscopic anomaly phenomena with 16 different types of anomalies were investigated. Among them, anomalies pertaining to groundwater and small animals dominated. The details of the recorded macroscopic phenomena are listed in Table 1. All of the anomalies listed in the Table 1 were witnessed or felt by at least three people, and we also conducted some additional surveys (Che et al. 2003).

These macroscopic phenomena were found in the cities of Xichang, Puge, Mianning, Yuexi, Muli, Xide, Ningnan, Huidong, and Qiaojia (Fig. 2). The following sections describe some of the typical anomalies that occurred during May and June in 2002 (additional information can be found in Table S1).

### 3.1.1 Animal Anomalies

1. The creeping of thousands of lucid worms on June 2nd, 2002, in the village of Chang. They had twisted into a group that made them appear to be long snakes crawling out from the underground. The lucid worms had approximate lengths of 1 cm with body diameters of 1–2 mm. In addition, the individual worms died quickly when they broke away from the group. The local residents told us that they had never seen such a swarm, let alone a group migration. When we sent this type of swarm to the Biological Research Institute in Xichang, they were unable to provide us with the exact name or tell us the habit of the swarm. In addition, they were found in the town of Huangshui of Xichang and in the towns of Puji and Qiaowo of Puge County in the following days (Che et al. 2003; Mu 2008).

2. The jumping of large numbers of fish in Qionghai Lake on June 10th, 2002. Qionghai Lake is the second-largest freshwater lake in Sichuan

Table 1

*Macroscopic anomaly phenomena that occurred from May to June of 2002 in the Xichang area*

| Types of anomalies | Specific behaviors of the anomalies | Number of anomalies | Reliability (*) |
|---|---|---|---|
| Groundwater | Spring drying up | 4 | A |
| | Artesian pressure or gushing in deep well | 3 | A |
| | Sharp decline in the water level | 2 | B |
| | Color change | 19 | B–C |
| | Smell and chemical compositional change | 7 | B–C |
| | Rise in the water level | 11 | A–B |
| Animals | Mice migration or running around | 13 | B |
| | Swarming migration of ants | 3 | B |
| | Crawling of group of snakes | 3 | B |
| | Creeping of lucid worms | 9 | A–B |
| | Jumping fish in Qionghai Lake and in fish ponds | 4 | A |
| | Groups of swallows staying on an electric wire throughout the night | 2 | A |
| | Pigeons breaking through nylon nets and flying out | 1 | B |
| | Cockroach suicide in groups | 1 | A |
| Others | Ground movements detected by humans | 1 | B |
| Total | 16 | 83 | |

* We divided the reliabilities of the anomalies into three levels: A, B, and C. A level of A means that those anomalies were witnessed by the authors and that the local inhabitants had never observed them before. A level of B means that those anomalies were witnessed or felt by many local people. A level of C means that the reliability of those anomalies could not be confirmed and that they may have been caused by the psychological effects of the possibility of future earthquakes

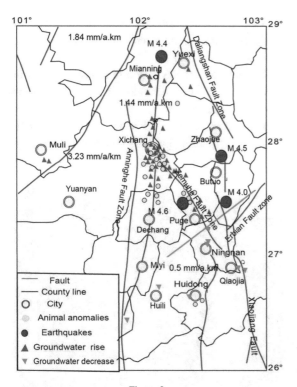

Figure 2
Distribution of macroscopic anomalies in the Xichang area

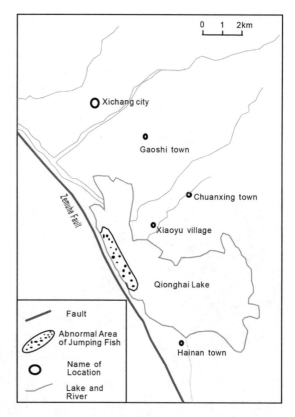

Figure 3
Locations of abnormal areas exhibiting jumping fish

province and is located to the southeast of Xichang at a distance of 5–10 km. The lake has a water area of nearly 31 km², and its western boundary is distributed along the Zemuhe fault (Fig. 3). In the afternoon of June 10th, 2002, many fish floated to the surface of the water with their heads pointed upwards on the water surface resembling a state of lacking oxygen. At 10:00 pm that night, fishermen discovered that tens of fish had begun to jump out of the water, and more fish (i.e., thousands) jumped out of the lake in total over the subsequent period. The height of their jumps reached as high as two meters. When a boat went across the area, more than 100 kg of fish jumped into the boat. The four fishermen on the boat were consequently sticky due to the fish. The area of abnormal jumping is located on the west bank of Qionghai Lake. We could still see some fish jumping after 30 min, which is when we arrived at the location. The anomaly area has a length of 2–3 km with a width of 200–300 m (Fig. 3). Such a phenomenon has never occurred before. It seems that this anomaly

is related with the tectonic activity along the Zemuhe fault, because the abnormal phenomenon occurred along the Zemuhe fault Zone (Fig. 3).

### 3.1.2 Groundwater Anomalies

The groundwater anomalies were the most prominent phenomena in our field survey, as 57 visible groundwater anomalies were identified. They exhibited multiple types of changes in the groundwater characteristics, such as changes in the water level, temperature, color and turbidity, drying up, or sharp decreases of the flow rate in springs, and demonstrations of artesian pressure or gushing in wells. Here, we describe several typical macroscopic groundwater anomalies.

1. The groundwater demonstrating artesian pressure in an abandoned well at the Erqing Supply and Marketing Company in the city of Xichang. The well has a depth of 100 m, and groundwater had

begun to flow out beginning on May 17th, 2002 (Fig. 4a).

2. The lifting of two well caps at a water supply plant in the town of Xiaomiao in Xichang by gushing groundwater observed on May 31st, 2002, at 5:00–5:10 pm (Fig. 4b, c). The caps were made of steel with diameters of 40 cm and thicknesses of 5 mm. When the water gushed out of the well, it contained some bubbles. After the gushing had stopped, the groundwater level in the well fluctuated several times.

3. The decrease of the flow rate in the spring in the village of Shisun, Ningnan County. The groundwater collected in a spring pool that was constructed prior to 1949 at the location of the spring. The spring pool is a rectangle with a length

Figure 4
Typical macroscopic groundwater anomalies in the Xichang area. **a** Groundwater in an abandoned well with artesian pressure; **b, c** two well caps lifted by gushing groundwater; **d** a spring pool becoming dried up at the village of Shixun

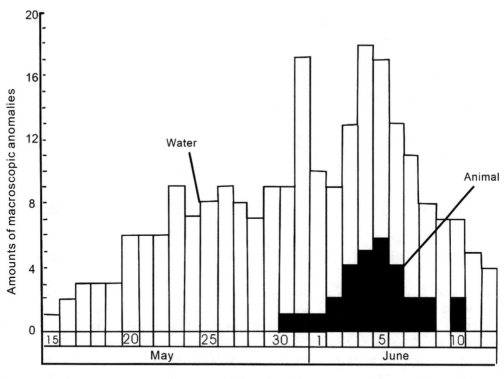

Figure 5
Anomaly frequency variations with time

of 1.2 m and a width of 1 m. The depth of the spring pool is larger than 1 m. More than three hundred people rely on it for drinking and agricultural irrigation. In the past few decades, the spring pool was always filled with water that discharged from an outlet on the south side of the spring pool. However, the flow rate had begun to decrease since March 2002, after which it decreased sharply beginning on May 20th, 2002, and was cut off on May 30th, 2002 (Fig. 4d).

4. The dwindling of water in a karst cave and changes in the turbidity in the village of Dongfeng, Ningnan County. This karst cave was expanded from a small spring. Now, it has a diameter of 1 m and a depth of 20 m. In the past few decades, water seepage had occurred everywhere within the cave; therefore, the cave had a rather large flow rate. The water in the cave is also clean, and tens of acres of terraces get their water supply from the cave. However, the water became turbid and less water spilled out from the cave beginning on June 2nd, 2002. When we arrived on

June 8th, 2002, the brim of the cave was almost dry and exhibited a very small flow rate.

### 3.1.3 Spatial and Temporal Characteristics of the Macroscopic Anomalies

The number of macroscopic anomaly phenomena reported varied daily during May and June of 2002 (Fig. 5).

Both the high-frequency macroscopic groundwater and animal anomalies occurred during the late May and the early June. The consistencies of the two types of anomalies indicate that they are reliable and can represent indicators of tectonic activity.

Spatially, the distribution of the macroscopic phenomena also shows some regular characteristics. Most of the anomalies (approximately 93%) occurred along the Anninghe and Zemuhe faults, and more than half of them were located in the area of intersection of the two faults (Fig. 2). Most of the animal anomalies occurred in the area of intersection, and no obvious boundary could be found. However,

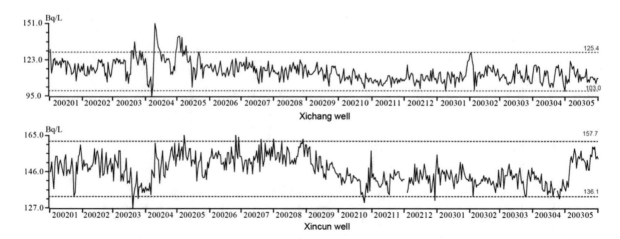

Figure 6
Radon anomalies in the Xichang well and in the Xincun well. The *dashed lines* represent two standard deviations of the background variation

an obvious difference in the changes of the regional groundwater level/flow rates was observed between the northwest segment of the Anninghe fault zone and the southeast segment of the Zemuhe fault (i.e., the groundwater level was elevated in the northwest part and declined in the southeast part on the border somewhere between Puge County and Ningnan County) (Fig. 2). The features of these groundwater level changes may suggest that the stress/strain status was different in those areas.

### 3.2. Microscopic Anomalies

In addition to the large-scale macroscopic anomalies, groundwater anomaly monitoring wells (i.e., collecting water level and water radon content data), fault gas measurements ($CO_2$), apparent resistivities, and crustal deformation measurements also synchronously exhibited anomalies.

### 3.2.1 Groundwater

The groundwater level showed increased anomalies in the northern part of the Anninghe fault zone (the Chuan 03, Chuan 05, and Chuan 18 wells) and decreased anomalies in the southern Zemuhe and northern Xiaojiang fault zones[1] (Cheng 2003). Radon concentrations in the Xichang well showed large

---

[1] From the Bulletin of Short-Term Earthquake Prediction in the Xichang area, 2002.6.22.

fluctuations that exceeded two standard deviations of the background level during the period from March to May of 2002. In addition, radon concentrations in the Xincun well showed low levels below two standard deviations during March 2002 (Fig. 6). In addition, the radon concentrations in the Shagou well showed a sustained decrease that exceeded two standard deviations during October 2001 and February 2002 (Fig. 7).

### 3.2.2 Fault Gas Discharge

Measurements of the $CO_2$ gas discharge from a fault zone have been considered as an effective way to detect earthquake precursors. In Mianning, the $CO_2$ gas discharge from the fault zone showed sustained rising anomalies since April 2001 that have been maintained at a high level since then (Fig. 8; Guan 2003).

### 3.2.3 Crustal Deformation (short-level data)

Short-level measurements in Tangjiaping of Ningnan County began to show a significant decrease beginning in 1997, but they started to increase after May 2001 (Fig. 9; Cheng 2003). The significant changes of these short-level measurements may indicate large changes of the stress state in the Zemuhe fault zone.

### 3.2.4 Apparent Electrical Resistivity

Apparent electrical resistivity measurements at the Mianning station showed a sustained decrease since

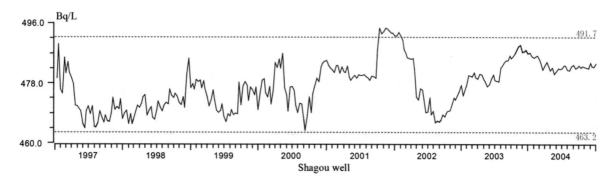

Figure 7
Radon anomalies in the Shagou well, located in the city of Panzhihua. The *dashed lines* represent two standard deviations of the background variation

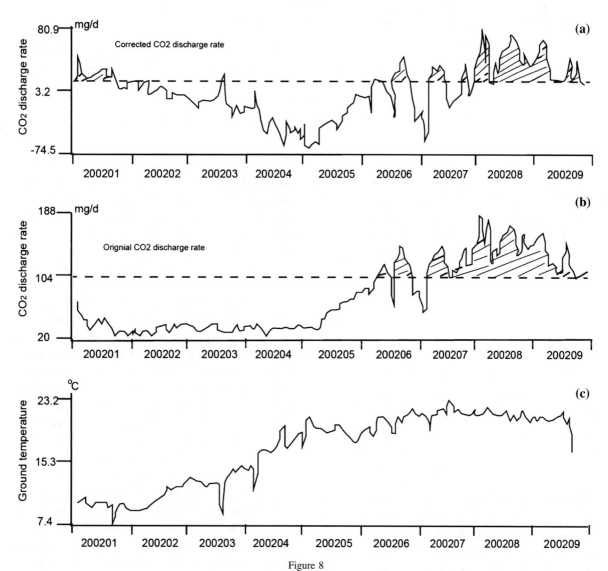

Figure 8
CO2 discharge rate anomalies at the Mianning station. **a** Corrected CO2 discharge rates over time; **b** original CO2 discharge rates; **c** ground temperature data (after Guan 2003)

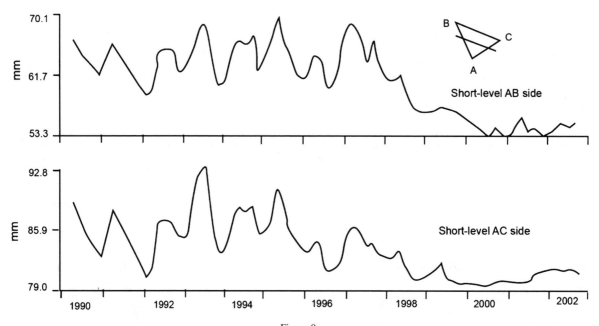

Figure 9
Short-level anomalies at the Tangjiaping station, Ningnan (after Cheng 2003)

1987, but the decreasing tendency stopped in the middle of 2001 and began to increase soon after (Fig. 10; Cheng 2003). Thus, the tendency of the apparent electrical resistivity was similar to that of the short-level records, and it may also indicate changes of the regional tectonic stress in the Zemuhe fault zone.

## 4. Discussion

### 4.1. The relationship between earthquakes and large-scale macroscopic anomalies

The Anninghe–Zemuhe–Xiaojiang Fault Zone is considered as one of the strongest seismically active belts in China, as many large earthquakes have occurred along the fault zone. Since 2001, a series of earthquakes have occurred that have disrupted the seismic quiescence in the Xichang area. In addition, large-scale macroscopic anomalies, which are considered as significant short-term precursor signals (Tributsch 1982; Soter 1999; Che et al. 2003), have also occurred since March 2002. Microscopic anomalies, such as groundwater level, groundwater chemistry, crustal deformation, and electric resistivity

anomalies, have occurred synchronously. Furthermore, the anomalies that occurred in the Xichang area are somewhat similar to those reported for the Haicheng earthquake: (1) large-scale macroscopic animal and groundwater anomalies were reported in the area; (2) small and medium earthquake activity also occurred in the area; and (3) the foreshock was regarded as an important factor in the successful prediction of the Haicheng earthquake (Wang et al. 2006). These previous experiences tell us that a strong earthquake is likely to occur after the occurrence of these large-scale anomalies. Thus, many researchers believe that a strong earthquake with a magnitude of 6 could occur after the occurrence of such macroscopic anomalies (Che et al. 2003). However, no strong earthquakes have occurred in this area, although many years have passed.

This lesson tells us that there is no direct relationship between an earthquake and these anomalies. In most cases, the large number of anomalies, including microscopic and macroscopic anomalies, is caused by tectonic activity. Earthquakes are only one representation of these tectonic activities. Accordingly, the relationship between these anomalies and an earthquake is "accompanying" and not "cause and

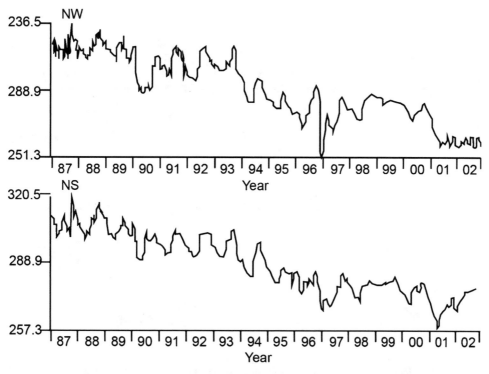

Figure 10
Apparent resistivity anomalies at the Mianning station (after Cheng 2003)

effect" in most cases. This means that these anomalies could occur as a consequence of tectonic activity, but they may occur independently of an earthquake.

## 4.2. Evidence of Macroscopic Anomalies Caused by Tectonic Activity

There are several facts that may support the conclusion that these macroscopic anomalies are caused by tectonic activity and that they are not earthquake precursors.

First, most of the anomalies were distributed along the Anninghe and Zemuhe faults, and half of them were located in the area of intersection of the two faults (Fig. 1). It is apparent that they were controlled by the tectonic setting. In addition, the distribution and behaviors of the distinct groundwater anomalies indicate a different stress/strain state along the different parts of the Zemuhe fault. As we know, the groundwater level will rise when an aquifer suffers from compression-like deformation and will drop in an area of extensional deformation. Accordingly, the macroscopic groundwater anomalies

indicate that compression-like deformation is dominant in northern Puge County, while a reverse displacement of extensional deformation is dominant in southern Ningnan County. This is also coincident with groundwater monitoring records in this fault zone: the wells along the northern Zemuhe fault zone showed an increase of the water level, whereas the groundwater level showed a decrease along the southern Zemuhe and northern Xiaojiang fault zones (Cheng 2003).

Why did the groundwater anomalies behave differently? According to the study conducted by Li et al. (2003), the Zemuhe fault zone can be divided into two different segments bounded near Puge County: the Xichang–Puge segment and the Puge–Ningnan–Qiaojia segment, both of which show different fault slippage rates. This is in agreement with the vertical deformation rate profile between Luohu and Yilang, which shows dissimilar vertical deformation rates between the Xichang–Puge and Puge–Ningnan segments (Zhang et al. 2003a, b; Fig. 11). An analysis of seismicity parameters using data from June 1976 to June 2001 shows that the

117

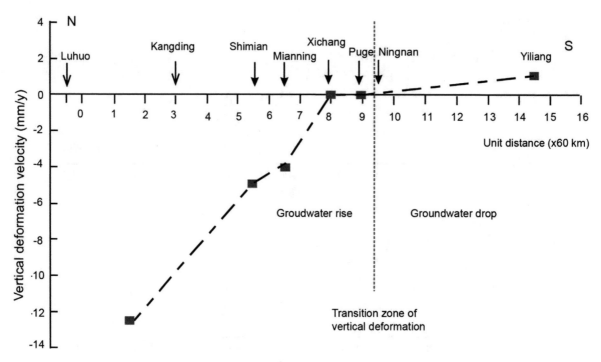

Figure 11
Vertical deformation rates along Luohuo–Yiliang profile (data from Zhang 2003)

Xichang–Puge segment of the Zemuhe Fault zone exhibited very low seismicity under a low stress level and that the Ningnan segment of the Zemuhe fault zone showed frequent seismicity and fault creep under a moderate stress level (Yi et al. 2004). Furthermore, the Erbian fault, which is one of the sub-faults of the Huayingshan Fault zone, intersects with the Zemuhe Fault Zone between Puge County and Ningnan County. This may also lead to the different distribution of groundwater anomaly behaviors in the northern and southern parts of the Zemuhe Fault zone. All of these studies indicate a different stress state within the northern and southern parts of the Zemuhe fault zone, which is coincident with the distribution of groundwater anomalies.

Second, seismo-geological and geodetic studies showed that movements along the Anninghe and Zemuhe faults were characterized by sinistral strike-slip motion with recently stronger compressional stresses in the Mianning–Xichang area (Zhang 2003; Yi et al. 2004; Zhang et al. 2003a, b), which correspond to the movement of the Tibetan Plateau toward the southeast, which results in the clockwise rotation of the Chuandian block (Shen et al. 2005;

Zhang et al. 2003a, b). This is also consistent with the distribution of the groundwater anomalies. The short-level and short-baseline measurements showed a movement of right-lateral rotation with a falling hanging wall near the area of Tangjiaping, Ningnan County, along the Zemuhe fault during 2002 (Fig. 9) (Du et al. 2002), which may be evidence of extensional deformation. The measurements also demonstrated an increase in strong tectonic activity along the eastern boundary of the Chuandian block. Such short-level changes in the observed anomalies may indicate changes in the regional stress/strain state and may be responsible for the distribution of groundwater anomalies.

Furthermore, six earthquakes with magnitudes ranging from $M_L 2$ to $M_L 4$ occurred over the span from May 23rd to June 21st, the period of which is in agreement with the occurrence of the macroscopic anomalies. Therefore, this also seems to confirm the tectonic nature of the macroscopic anomaly phenomena.

All of this evidence indicates that the macroscopic anomalies that occurred in 2002 were caused by tectonic activity along faults. From the overall characteristics of the tectonic activity, we can find that the tectonic activity has a distinct dividing line between Puge County and Ningnan County. The northern part, which is the Mianning–Xichang–Puge area, is dominated by a compressional tectonic environment and is accompanied by the accumulation of small amounts of energy. The southern part, which includes the Ningnan and Qiaojia areas, is dominated by an extensional tectonic environment, which plays an important role in the release of energy. Seismic activity and changes of the groundwater level are the main methods by which energy is released. In addition, we can deduce that the energy accumulation in the north and the energy release in the south represent a constant process of energy balance and transference. During this process, small-to-moderate earthquakes may always occur in the area of southern Puge County and Butuo County, which are controlled by extensional tectonics. In addition, when the energy released within the extensional area is not enough for the amount that is transferred from the compressional area, the compressional area may suffer from an increased risk of medium or even strong earthquakes.

## 5. Conclusions

We documented large-scale anomalies, particularly macroscopic anomalies, that occurred from May to June of 2002 in the Xichang area, southwestern Sichuan Province, China. Animal, groundwater, apparent electrical resistivity, seismic activity and geodetic anomalies were found during our field work. According to the investigation, we discovered the following conclusions.

1. Temporally, a good agreement between the peak frequencies of macroscopic groundwater and animal anomalies was reached during the late May and the early June, which was coincident with the level of increasing seismic activity in this area.
2. Spatially, the distributions of the groundwater and animal anomalies were closely related with the tectonic setting. The distinct differences of the behaviors of the groundwater anomalies that were divided between Puge County and Ningnan

County indicate different levels of activity along different segments of the Zemuhe fault.
3. Macroscopic phenomena will occur before both strong tectonic activity and strong earthquakes. However, whether a strong earthquake will occur depends upon the amount of energy accumulated within the fault.

## Acknowledgements

We greatly thank Cheng Wanzheng, Wen Xueze, Guan Zhijun, Zhang Yongjiu, An Mingzhi, Liu Dean, and Hu Fangliang for their help during the field work. This work is supported by the National Natural Science Foundation of China (41602266, U1602233), the Fundamental Research Funds for the Central Universities (2652015316), and the Beijing Talents Funds (2016000020124G110).

## References

Bakun, W. H., Aagaard, B., Dost, B., Ellsworth, W. L., Hardebeck, J. L., Harris, R. A., et al. (2005). Implications for prediction and hazard assessment from the 2004 Parkfield earthquake. *Nature, 437*(7061), 969.

Che, Y. T., & Liu, C. L. (2008). Opinions about earthquake prediction after Wenchuan earthquake. *Recent Development in World Seismology, 10,* 1–6.

Che, Y., Wang, G., Liu, C., & Zhu, Z. (2003). Large scale macro-abnormal phenomena appearing 2002 in Lianshan predecture region and its analysis. *Earthquake Research in Sichuan* (3), 14–19.

Che, Y., & Yu, J. (2006). *Underground fluid and earthquake.* Beijing: China Meteorological Press.

Cheng, W. (2003). Some precursory phenomena predicting the tendency of strong earthquakes along the seismic belts of Anninghe–Zemuhe–Xiaojiang and application thoughts. *Earthquake Research in Sichuan.*

Du, F., Cheng, W., Guan, Z., & Li, G. (2002). *2003 year's research report of the trend of earthquakes in Sichuan Province.* Yingxiu: Sichuan Earthquake Administration.

Geller, R. J. (1997). Earthquake prediction: A critical review. *Geophysical Journal International, 131*(3), 425–450.

Gold, T., & Soter, S. (1984). Fluid ascent through the solid lithosphere and its relation to earthquakes. *Pure and Applied Geophysics, 122*(2), 492–530.

Guan, Z. (2003). Tracking study of recent water Chemistry anomalies in key monitoring area of Sichuan and the boundary between Sichuan and Yunan. *Earthquake Research in Sichuan, 107*(2), 45–48.

Hartmann, J., & Levy, J. K. (2005). Hydrogeological and gasgeochemical earthquake precursors—A review for application. *Natural Hazards, 34*(3), 279–304.

Ingebritsen, S. E., & Manga, M. (2014). Earthquakes: Hydrogeochemical precursors [News and Views]. *Nature Geoscience*. doi:10.1038/ngeo2261.

King, C. Y., Azuma, S., Ohno, M., Asai, Y., He, P., Kitagawa, Y., et al. (2000). In search of earthquake precursors in the water-level data of 16 closely clustered wells at Tono, Japan. *Geophysical Journal International, 143*(2), 469–477.

Liu, C. L., Che, Y. T., & Wang, G. C. (2004). Duality of large-scale macro-anomalies and their implication to earthquake prediction. *Seismological and Geology, 26,* 340–346.

Lomnitz, C. (1994). *Fundamentals of earthquake prediction.* New York: Wiley.

Manga, M., & Wang, C.-Y. (2015). Earthquake hydrology. In G. Schubert (Ed.), *Treatise on geophysics* (2nd ed., Vol. 4, pp. 305–328). Amsterdam: Elsevier.

Matsumoto, N., & Koizumi, N. (2013). Recent hydrological and geochemical research for earthquake prediction in Japan. *Natural Hazards, 69*(2), 1247–1260.

Mu, Y. (2008). The practice and thinking of short-term earthquake forecast for southwest Sichuan and its adjacent region. *Recent Developments in World Seismology, 5,* 12–22.

Peresan, A., Kossobokov, V., Romashkova, L., & Panza, G. (2005). Intermediate-term middle-range earthquake predictions in Italy: a review. *Earth-Science Reviews, 69*(1), 97–132.

Rikitake, T. (1994). Nature of macro-anomaly precursory to an earthquake. *Translated World Seismology, 42*(2), 149–163.

Roeloffs, E. A. (1988). Hydrologic precursors to earthquakes: A review. *Pure and Applied Geophysics, 126,* 177–209.

Scholz, C. H., Sykes, L. R., & Aggarwal, Y. P. (1973). Earthquake prediction: A physical basis. *Science, 181*(4102), 803–810.

Shen, Z.-K., Lü, J., Wang, M., & Bürgmann, R. (2005). Contemporary crustal deformation around the southeast borderland of the Tibetan Plateau. *Journal of Geophysical Research, 110*(B11), B11409.

Skelton, A., Andrén, M., Kristmannsdottir, H., Stockmann, G., Morth, C.-M., Sveinbjornsdottir, A., et al. (2014). Changes in groundwater chemistry before two consecutive earthquakes in Iceland. [Letter]. *Nature Geoscience, 7,* 752–756. doi:10.1038/ngeo2250.

Soter, S. (1999). Macroscopic seismic anomalies and submarine pockmarks in the Corinth-Patras rift, Greece. *Tectonophysics, 308*(1), 275–290.

Toutain, J. P., & Baubron, J. C. (1999). Gas geochemistry and seismotectonics: A review. *Tectonophysics, 304*(1–2), 1–27.

Tributsch, H. (1982). *When the snakes awake: Animals and earthquake prediction.* Cambridge: MIT Press.

Wallace, R. E., & Teng, T. L. (1980). Prediction of the Sungpan–Pingwu earthquakes, August 1976. *Bulletin of the Seismological Society of America, 70*(4), 1199–1223.

Wang, K., Chen, Q.-F., Sun, S., & Wang, A. (2006). Predicting the 1975 Haicheng earthquake. *Bulletin of the Seismological Society of America, 96*(3), 757–795.

Wang, C. Y., & Manga, M. (2010). *Earthquakes and Water.* Berlin: Springer.

Wang, G., & Shen, Z. (2010). Seismic groundwater monitoring and earthquake prediction. *Chinese Journal of Nature, 32*(2), 90–93.

Wen, X., & Cheng, N. (1992). Study on seismic potential of the key region of earthquake monitoring and defending of Sichuan. *Earthquake Research in Sichuan, 1,* 1–25.

Woith, H. (2015). Radon earthquake precursor: A short review. *The European Physical Journal Special Topics, 224*(4), 611–627.

Wyss, M. (1997). Cannot earthquakes be predicted? *Science, 278*(5337), 487–490.

Yan, Y. K. (1997). Are earthquakes predictable? *Geophysical Journal International, 131*(3), 505–525.

Yi, G.-X., Wen, X.-Z., Fan, J., & Wang, S.-W. (2004). Assessing current faulting behaviors and seismic risk of the Anninghe–Zemuhe fault zone from seismicity parameters. *Acta Seismologica Sinica, 17*(3), 322–333.

Zhang, C. (2003). Research on the recent risk zone of strong earthquake in the Xianshuihe and Xiaojiang fault of Sichuan and Yunnan area. *Seismological consideration of 2003* (pp. 60-73): Institute of Geology, China Earthquake Adminstration.

Zhang, P., Deng, Q., Zhang, G., Ma, J., Gan, W., Min, W., et al. (2003a). Active tectonic blocks and strong earthquakes in the continent of China. *Science in China, Series D: Earth Sciences, 46,* 13–24.

Zhang, G., & Fu, Z. (1981). Some features of medium-and short-term anomalies before great earthquakes. *Maurice Ewing Series 4: Earthquake Prediction*, 497-509.

Zhang, X., Jiang, Z., Wang, Q., Wang, S., & Zhang, X. (2003b). Recent crust motion features in Sichuan–Yunnan and their relationship to strong earthquake. *Journal of Geodesy and Geodynamic, 23,* 35–41.

Zhang, H., & Li, K. (1977). *Characteristic of precursory groundwater dynamic in Tangshan earthquake (Precursor of Tangshan earthquake).* Beijing: Seismological Press.

Zhu, F., & Wu, G. (1982). *Haicheng earthquake in 1975.* Beijing (in Chinese): Seismological Press.

(Received February 25, 2017, revised June 14, 2017, accepted June 21, 2017)

Pure Appl. Geophys.
© 2017 Springer International Publishing AG
DOI 10.1007/s00024-017-1626-7

❙ **Pure and Applied Geophysics**

# Experimental Study of Thermal Field Evolution in the Short-Impending Stage Before Earthquakes

YAQIONG REN,[1,2] JIN MA,[2] PEIXUN LIU,[2] and SHUNYUN CHEN[2]

*Abstract*—Phenomena at critical points are vital for identifying the short-impending stage prior to earthquakes. The peak stress is a critical point when stress is converted from predominantly accumulation to predominantly release. We call the duration between the peak stress and instability "the meta-instability stage", which refers to the short-impending stage of earthquakes. The meta-instability stage consists of a steady releasing quasi-static stage and an accelerated releasing quasi-dynamic stage. The turning point of the above two stages is the remaining critical point. To identify the two critical points in the field, it is necessary to study the characteristic phenomena of various physical fields in the meta-instability stage in the laboratory, and the strain and displacement variations were studied. Considering that stress and relative displacement can be detected by thermal variations and peculiarities in the full-field observations, we employed a cooled thermal infrared imaging system to record thermal variations in the meta-instability stage of stick slip events generated along a simulated, precut planer strike slip fault in a granodiorite block on a horizontally bilateral servo-controlled press machine. The experimental results demonstrate the following: (1) a large area of decreasing temperatures in wall rocks and increasing temperatures in sporadic sections of the fault indicate entrance into the meta-instability stage. (2) The rapid expansion of regions of increasing temperatures on the fault and the enhancement of temperature increase amplitude correspond to the turning point from the quasi-static stage to the quasi-dynamic stage. Our results reveal thermal indicators for the critical points prior to earthquakes that provide clues for identifying the short-impending stage of earthquakes.

**Key words:** Stick slip, meta-instability, quasi-static, quasi-dynamic, temperature variation, thermal field.

## 1. Introduction

Exploring anomalies in the short-impending stage prior to earthquakes is an important way to reduce losses caused by earthquake disasters; however, a great number of earthquakes have occurred without observed premonitory changes. Many studies have been undertaken to identify the arrival of earthquakes early in the short-impending stage.

The $M_s$ 7.3 earthquake in Haicheng, China, that occurred on February 4, 1975, was predicted in advance, and disaster mitigation measures were adopted that attracted widespread attention both at home and abroad. The basis of prediction was established from more than 500 foreshocks that occurred within 4 days and the apparent increase in foreshock activity 26 h prior to the main shock(Wu et al. 1976). Unfortunately, no foreshocks occurred before the $M_s$ 7.8 Tangshan earthquake on July 28, 1976, and there was even a seismic quiescence in North China half a year prior to the main shock (Mei 1986). Statistics show that only a small number of large earthquakes occur with foreshocks (Jordan et al. 2011). Even with obvious foreshocks identified by retrospective studies after earthquakes, foreshocks are difficult to distinguish from general small earthquakes prior to the earthquake.

Zhang and Fu (1990) suggested that earthquakes do not occur at peak stress but at a certain point during a stage of slip weakening after peak stress. Slip weakening could be the important physical mechanism of entering the short-impending stage and brought about a variety of short-term precursors.

Yin et al. (1995, 2000, 2008) proposed a method of predicting earthquakes using the Load-Unload Response Ratio (LURR), which is based on the

---
[1] Key Laboratory of Seismic Observation and Geophysical Imaging, Institute of Geophysics, China Earthquake Administration, No.5 Minzudaxuenan Rd, Beijing 100081, China.
[2] State Key Laboratory of Earthquake Dynamics, Institute of Geology, China Earthquake Administration, No.1 Huayan Rd, Beijing 100029, China. E-mail: majin@ies.ac.cn

nonlinearity of the stress–displacement curve when the damage process begins. The LURR value represents the regional extent of damage, and the maximum value corresponds to peak stress. Earthquakes can be predicted to occur several months after peak stress has been captured. The theory and calculation of LURR have clear physical meanings and have received extensive attention and support.

Das and Scholz (1981) and Dieterich (1986) discussed seismic nucleation models from the perspectives of rupture propagation and frictional sliding, respectively. Both models assumed that there is a stable sliding portion on the fault plane, and emphasized that when the slip distance expands to a critical length, stable slip transitions to accelerated slip until instability occurs. The process occurring from the onset of stable sliding to unstable sliding is called the nucleation process. Dieterich (1992) modeled the spontaneous nucleation of slip instabilities using rate- and state-dependent constitutive laws, and noted that the scaling of $D_c$ (the characteristic slip distance), which varies with surface roughness and gouge particle size, may possibly be the key factor that determines whether the nucleation process can be observed in the field.

Ohnaka et al. (1987) studied the constitutive relations between shear stress, slip velocity, slip acceleration and displacement in the breakdown zone during stick–slip shear failures of rock. The nucleation process prior to instability was revealed, and the size-scale dependence of the above physical parameters was discussed to extend the laboratory results to natural faults. Similar results were also obtained in larger scale rock samples (Okubo and Dieterich 1984).

The earthquake nucleation model received much attention, because it was regarded as among the most important advances in seismology in the 1990s. Many studies have been conducted that have examined, for example, the effects of normal stress, the loading histories and fault uniformity on critical length, the relationship between the nucleation process and ultimate earthquake magnitude, and the seismic validation of earthquake nucleation in the field (Kato et al. 1992; Ellsworth and Beroza 1995; Kato and Hirasawa 1996; Ma et al. 2003; Thompson et al. 2009; Fang et al. 2010; McLaskey and Lockner 2014; Hasegawa and Yoshida 2015; Yasuo et al. 2015).

Regarding field practices, a predicted $M$ 6.4 earthquake in 2004 in southern California never occurred, and the California Northridge earthquake in 1994 and Japanese Kobe earthquake in 1995 occurred without cues (Cyranoski 2004). Claims regarding the unpredictability of earthquakes had prevailed (Geller et al. 1997). However, the silent slip events and non-volcanic tremors in the Cascadia subduction zone were found to repeat at 13–16 month intervals (Rogers and Dragert 2003). The widespread existence of those phenomena in the subduction zones and boundary regions of continental plates has ignited hopes for earthquake prediction. However, subsequent observational studies have shown that slow earthquakes have different forms in different regions, and among repeated slow earthquakes in the same area (Ida 2010; Melgar et al. 2016). The complexity of earthquake generation demands various types of studies to understand the mechanism of the short-impending stage.

Because the frictional heat generated during earthquake faulting is thought to be the largest part (80–90%) of the total seismic energy budget (Kano et al. 2006), it might be possible to find clues from variations in thermal fields when entering the short-impending stage. Studies have shown that the causes of thermal variations during rock deformation not only include the commonly referenced frictional sliding but also include lesser-known stress changes (Liu et al. 2004b; Wu et al. 2006; Chen et al. 2009, 2015). Beam bending tests and cyclic loading tests of steel and granodiorite (Liu et al. 2004a, b; Chen et al. 2009, 2015) have been carried out to investigate the relationship between elastic deformations and thermal variations. The results show that the surface temperatures of rock increase under loading conditions and decrease under unloading conditions linearly. The deformations of complex structures such as en-echelon faults are also accompanied by temperature increases in compressional jog areas and temperature decreases in extensional jog areas (Ma et al. 2007, 2010). According to the first and second laws of thermodynamics, it can be deduced that a linear relationship exists between changes in temperature and stress that depend on coefficients of thermal expansion, specific heat and density of materials (Liu et al. 2004b; Chen et al. 2009, 2015).

The thermal evolution of stick slips on bending faults (Ma et al. 2012) shows that there are different temperature variations during the two deformation stages before and after peak stress: in the prior stage, temperatures in the fault and wall rocks both increase, whereas in the last stage, temperatures in the faults increase, but temperature in the wall rocks decreases. This result indicates that thermal parameters can be used to identify peak stress.

The above thermal observations were obtained using platinum resistance thermometers placed on the rock surfaces. The spatial resolutions of the results strongly depended on the distribution of the measuring points. Due to the limited spatial resolution, important changes could have been missed, resulting in a failure to achieve a full expression of the variations in the stress field. To obtain more specific characteristics of the different deformation stages, especially the short-impending stage, we used a thermal infrared imaging system in this study. The system is a full-field observational apparatus with high spatial resolution and continuous acquisition.

## 2. Experimental Design

### 2.1. Press and Sample

The experiment was conducted on a horizontally bilateral servo-controlled press machine (Fig. 1a). The press could be independently controlled in the X and Y directions through stress or displacement, the control frequency was 20 Hz, and the sampling frequency was 10 Hz. The loading capacity was 120 tons.

The 300 mm square and 50 mm thick granodiorite sample was sourced from Fang-shan County, Beijing, and is shown in Fig. 1b. A planar strike slip fault was precut diagonally through the sample. The roughness of the fault surface was ~10 μm. The rock near the fixed rams is called the fixed wall, and the rock near the moving rams is called the active wall.

### 2.2. Instrument and Data Processing

We used a full-field ImageIR 8820 system from Infra Tec (Dresden, Germany) to record the thermal infrared changes induced by deformation at a speed of 50 frames per second. The infrared camera with spectrum ranging from 8 to 14 μm was fixed 0.5 m above the sample. The minimum temperature resolution of the AD mode transform is 2.5 mK at room temperature, and the noise equivalent temperature difference of a single pixel is 25 mK. The length of a single pixel in this study is 0.345 mm. To ensure reliable thermal infrared observations, the experiment was conducted while avoiding lights and the movements of persons.

The temperature data recorded by the infrared thermal imaging system included not only information concerned with deformation but also environmental temperature variations and fluctuations caused by system noise and the random noise of

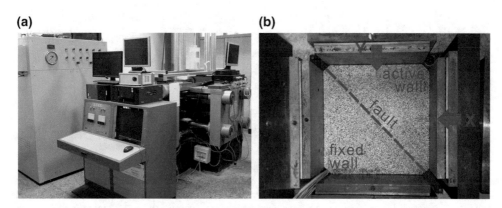

**(a)** **(b)**

Figure 1
Horizontally bilateral servo-controlled press machine (**a**) and sample design (**b**). **b** The *red dashed line* shows the precut fault, which divides the sample into two parts: the fixed wall and the active wall. The *red arrows* show the loading directions

the instrument itself. The environmental temperature variations were affected by room temperature, thermal radiation from the working apparatuses and human activities. The system noise of the instrument was primarily due to the inhomogeneity and temperature drift of the sensors, and the random noise was primarily due to the sensor's thermal and supply current noise. To obtain the thermal information related to deformation and improve the temperature resolution, three data processing procedures were applied. First, room temperature changes were removed by subtracting the temperature of a reference object (a portion of the unforced bottom of the press machine). Second, system noise originating from the sensor inhomogeneities and unevenly distributed environmental temperatures caused by the working apparatuses were reduced by spatially subtracting a reference image. System noise caused by temperature drift can be ignored given the camera's cooling system and short durations of the studied period. Lastly, random noise was decreased using the neighborhood average smoothing method (16 pixels × 16 pixels) spatially and the moving average smoothing method (window of 0.08 s and step of 0.02 s) in time. By employing those methods, the temperature resolution was improved from 25 mK to 2–3 mK. The stress sensitivity coefficient (thermal elastic coefficient) was 1 mK/MPa based on theoretical calculations and experimental observations (Chen et al. 2015). In this study, the smallest stress

drop for the stick slip exceeded 2 MPa. Thus, the stress changes were recorded at the then current temperature resolution.

## 3. Results

### 3.1. Experimental Procedures and Deformation Stages of a Stick–Slip Event

A series of stick slip experiments under different lateral pressures (applied in the $X$ direction) and loading rates were conducted to study the thermal variations prior to instability. Two stick slip experiments "EXP1" and "EXP2" with lateral pressures of 5 MPa and 7.5 MPa were analyzed for this article. A typical stress–time curve at a lateral pressure of 5 MPa with different loading rates is shown in Fig. 2a. Generally, the slower loading rate extended the duration of the stick slip events, and the greater lateral pressure increased the stress drops of the stick slip events.

The "EXP2-SS4" stick slip event under a lateral pressure of 7.5 MPa was selected for detailed analysis. The loading pattern of that experiment can be divided into two stages. (1) First, loading was synchronously applied in both the $X$ and $Y$ directions of the sample at the same loading rate of 20 kg/s to a constant 7.5 MPa. (2) The $Y$-direction load was then changed to displacement control with rates of 0.5 and 0.1 µm/s, whereas the $X$-direction load remained

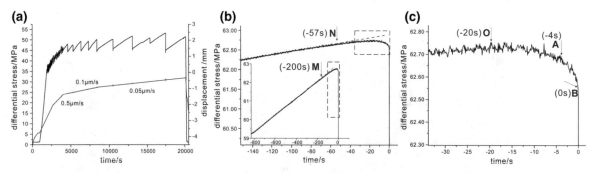

Figure 2

Typical stress- and displacement–time curves of the whole experimental procedure and deformation stages of a stick slip event: **a** typical stress- and displacement–time curves of the whole experimental procesdure. **b** An enlarged view of the *dashed rectangle* in the inset. **c** An enlarged view of the *dashed rectangle* in **b**. The *inset* in **b** shows the variations in differential stress over time throughout the entire stick–slip period. The meanings of each letter in the figure are as follows: *M* denotes the start of the deviation-from-linearity stage. *N* denotes the start of the strong deviation-from-linearity stage. *O* denotes the time of peak differential stress. *A* denotes the accelerating time of stress release. *B* denotes the time of instability constant. It should be noted that the sample was loaded at an equal displacement rate rather than an equal stress rate to produce repeated sequences of stick slip events without resets. As to the actual fault motion, it is more likely to be involved with both patterns.

The division of the deformation stages was mainly based on the differential stress vs. time curve obtained at the rear of the press ram. The inset in Fig. 2b shows the variations in differential stress over time during the entire stick–slip duration. It can be seen that the rock deformation was elastic at the initial stage, and the rock was damaged when the curve deviated from linearity at point *M*. Figure 2b shows an enlarged view of the dashed rectangle in the inset. The curve's deviation from linearity was more noticeable with the increasing degree of damage, and the onset of the strongly deviated-from-linearity stage (*NO*) occurred at the time marked by *N*. Figure 2c shows an enlarged view of the dashed rectangle in Fig. 2b. The time of peak differential stress is designated *O*, the accelerating time of stress release is designated *A*, and the time of instability is designated *B*. The stress prior to the peak point was mainly accumulated, which is referred to as "stick". After *O*, stress began to release until instability occurred, which is referred to as "slip". The instability thus did not occur suddenly but was preceded by a preparation stage that is referred to as the "meta-instability stage" (*OB*) (Ma et al. 2012). A notable characteristic of that stage is stress release, which indicates that some portions of the fault have begun to slip and that the stick slip event has entered a spontaneous instability stage. The meta-instability stage can be further subdivided into two stages, the quasi-static stage (*OA*) and the quasi-dynamic stage (*AB*), according to the stress release rate. The duration of this event was approximately 800 s, the quasi-static stage continued for 20 s, or 2.5% of the total duration, and the quasi-dynamic stage lasted for 4 s, or 0.5% of the total duration.

### 3.2. Identification of Deformation Stages Based on Thermal Variations

After following the data-processing procedures listed in 2.2, we obtained approximately 3000 frames

(60 s) of thermal increment images before and after the "EXP2-SS4" instability point. The relative temperature values were obtained by subtracting the temperatures at reference time N.

Figure 3a–d shows the representative thermal increment images taken in the different deformation stages of the "EXP2-SS4" stick slip event. Figure 3e shows the range of the afore-mentioned images (132 mm × 182 mm) on the sample and the image times in the differential stress vs. time curve. When the instability time was set to 0 s, the O and A times were −20 and −4 s, respectively. Figure 3a shows that the temperatures continually increased overall prior to the peak stress at *O*. At that stage, the fault and wall rocks deformed as a whole, and the location of the fault could not be distinguished in the thermal images. However, Fig. 3b–d shows a relative temperature increase along the fault and a relative temperature decrease in the wall rocks after the peak stress at O. In the quasi-static stage (Fig. 3b), the temperature of the wall rocks tended to decrease with increasing amplitude, which is an important thermal indicator of entering the meta-instability stage. In that stage, obvious positive temperature changes occurred only on sporadic sections on the left part of the fault. In the quasi-dynamic stage (Fig. 3c), the warming sections of the fault spread and connected with each other with higher warming amplitude. The location of the fault was easily discerned in the thermal images, and almost the entire fault was warmed 1 s prior to the instability. The rapid expansion of the regions of temperature increases on the fault and enhanced temperature increase amplitudes are important thermal indicators of entering the quasi-dynamic stage. In that stage, the temperature of the wall rocks still decreases, with a reducing amplitude. When instability occurred (Fig. 3d), the temperature of the fault rapidly increased and the temperature of the wall rocks decreased. The co-seismic temperature increase of the fault is 947 mK, and the co-seismic temperature decrease of the wall rocks is 30 mK. The temperature decrease was caused by a stress release of 24.8 MPa, and the gradient was 1.2 mK/MPa. The gradient was in accordance with theoretical results (Chen et al. 2015).

Figure 3 shows spatial images at given times and with less time scale information. A profile perpendicular to the fault (profile 1 in the inset of Fig. 4b)

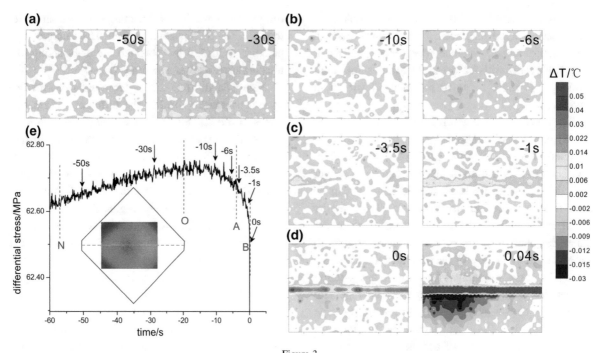

Figure 3

Thermal increment images of the different deformation stages (relative to time *N*): **a** the overall temperature increase in the stage of stress accumulation. **b** The temperatures decrease in most of the wall rocks in the quasi-static stage. **c** The temperature on the fault increase, and the sections of increasing temperature spread spatially in the quasi-dynamic stage. **d** The temperature on the fault increases rapidly and decreases in the wall rocks during and after the instability. **e** The locations of the above images on the sample and the times of each image in the differential stress vs. time curve

was made to obtain the continuous temperature changes of the fault and the wall rocks over time. Figure 4a shows a contour map of the temperature variations along profile 1. The horizontal axis represents the distance to the fault, and the negative and positive numbers correspond to the top and the bottom of profile 1, with zero as the location of the fault and the other numbers as the location of wall rocks. The vertical axis represents time, with every deformation stage defined by a differential stress vs. time curve marked by the side. Figure 4a shows that the temperature of the fault increased with the increasing temperature of the wall rocks before *O*, remained relatively high as the wall rocks cooled after *O*, and exhibited a distinct increase in amplitude when entering the quasi-dynamic stage. Thus, the temperature variations in every stage listed in Fig. 3 are representative. Because profile 1 represents only a small section of the sample, we calculated the average temperature across the entire fault and wall rocks, as shown in Fig. 4b. The calculation areas are

shown in the inset of Fig. 4b using red and blue rectangles, where the width of the fault area that we selected was 17 pixels, which was much larger than the width of the actual fault (less than one pixel), taking into account the stability of the data and the thermal conduction in the space. Thus, the temperature variations on the fault inevitably included the effects of stress changes in the wall rocks. The most obvious thermal field characteristic between the fault and the wall rocks was the opposite variation trend after entering the meta-instability stage, and that trend was intensified with the approach of the instability. Both figures demonstrate that the deformation stages of stick slip can be inferred from the evolution of thermal fields.

### 3.3. Spatial and Temporal Thermal Variations Along the Fault

A profile along the fault (profile 2 in the inset of Fig. 5c) was chosen for the temperature analysis on

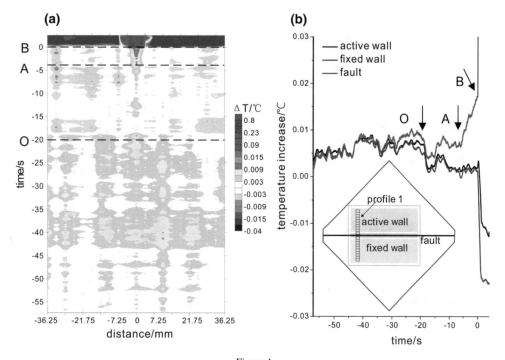

Figure 4
Temperature variations on the fault and wall rocks in the different deformation stages: **a** a contour map of the temperature variations of profile 1, with distance to the fault as the horizontal axis and time as the vertical axis. The fault is located at "0", and the remaining positions are wall rocks. **b** The average temperature variations of the entire fault and wall rocks with time. The vertical axis is the temperature increase relative to time N. O, A and B in the figures mark the deformation stage, which is defined by a differential stress vs. time curve. The *inset* of **b** shows the location of the profile and the areas used for calculate the average temperatures

different portions of the fault. The width of the profile was 1.7 mm and covered the entire fault and a small part of the wall rocks. Figure 5a shows a contour map of the temperature variations along the fault. The horizontal axis represents the distances to the start point of the profile. Regions with increasing temperatures slowly spread in the quasi-static stage but rapidly expanded along the fault in the quasi-dynamic stage. Two seconds prior to the instability, the temperatures of most regions of the fault increased, with the left end having the warmest regions. The contour map shown in Fig. 5b is similar to that in Fig. 5a but has a smaller time scale. It shows that the temperature of the entire fault increased 80 ms prior to the instability. That is a phenomenon of collaborative temperature increases, which is in accordance with previous results obtained using thermometers, strain gages and the digital speckle correlation method (Ma et al. 2012, 2014; Zhuo et al. 2013). At the same time, the heating rate increased over time

by orders of magnitude: 0.3, 3.6, and 5758 mK/s during the quasi-static stage, the quasi-dynamic stage, and the instability stage. That is, the temperatures and areas with increasing temperatures on the fault increased in the meta-instability stage.

To better describe the activity of the fault, the $A_F$ index was constructed, which considers increases in both the amplitudes and regions of temperature increments during the stick slip events:

$$A_F = \sum \Delta T \times \frac{N_{\Delta T}}{N_0}$$

A unit of $3 \times 3$ pixels was selected to calculate the average value in the fault zone ($9 \times 528$ pixels). The first item in the equation is the sum of the temperature increments of all the calculation units relative to time N. The second item in the equation is the ratio of the number of units with increasing temperature ($N_{\Delta T}$, $\Delta T > 0.002$ °C) to the total number of calculating units ($N_0$). The index can be used to

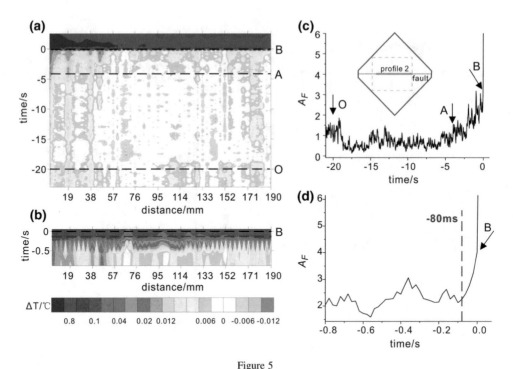

Figure 5
Temperature variations along the fault and the index of fault activity: **a** the contour map of temperature variations along the fault. The regions of increasing temperature spread slowly in the quasi-static stage, and rapidly in the quasi-dynamic stage. **b** A contour map with a smaller time scale, which shows a phenomenon of collaborative temperature increases 80 ms prior to the instability. **c** An obvious increase in the fault activity index after A. The *inset* of **c** shows the location of the profile. **d** An abrupt increase in the rate of the fault activity 80 ms prior to the instability

synthetically describe the fault activity, as shown in Fig. 5c. The activity was weak in the quasi-static stage and became enhanced after entering the quasi-dynamic stage. An abrupt increase in the rate of the fault activity 80 ms prior to the instability can be seen in Fig. 5d, which is in accordance with Fig. 5b that expressed growth both in amplitude and the warming regions and indicated imminent instability.

## 4. Discussion

### 4.1. Mechanisms of Thermal Variations During the Meta-Instability Stage

The analysis of the short-impending stage prior to earthquakes was based on the temporal relationship between anomalies and earthquakes. However, whether the instability will occur is determined by the stress state of the fault rather than time. As we previously mentioned, the period between the peak stress and the instability in a stick slip cycle is called

the meta-instability stage. The peak stress at $O$ corresponds to the critical point when the differential stress changes from accumulation to release, and the turning point at $A$ is the other critical point when the release of the fault transitions from quasi-static to quasi-dynamic. Both can be identified from a thermal field.

Before reaching the peak stress at $O$, the stress throughout the whole sample accumulated, and energy was stored in the wall rocks, causing overall temperature increases (Fig. 3a). When instability occurred, the stored strain energy was transformed into frictional kinetic energy and heat energy, which was manifested as a decrease in the temperature of the wall rocks due to the stress release and a temperature increase on the fault generated by sliding friction (Fig. 3d). The mechanisms for those temperature variations are relatively clear.

When entering the meta-instability stage, some sections of the fault begin to slip. The mechanism of the temperature decrease in the wall rocks is undoubtedly stress relaxation, whereas the mechanism of the temperature increase on the fault could be local fault slip or stress concentration due to stress redistribution in the wall rocks near the fault. To distinguish friction-generated heat from stress-caused heat, the relative displacement between the two walls of the fault needs to be analyzed. A larger relative displacement corresponds to a longer slide distance, which generates more heat along the fault. Digital images of the sample surface were obtained via the camera at a rate of 30 frames/s, and processed using a digital speckle correlation method (Zhuo et al. 2013). The fault relative displacement was measured using digital gauges that were placed perpendicular to the fault and with the same distances from the two endpoints to the fault. By calculating the displacements of the two endpoints, we can obtain the relative displacements parallel and perpendicular to the fault. In this study, 100 digital gauges that were 6.4 mm long were arranged at 1.7 mm intervals in the same area as the thermal images. The results were calculated using the GeoDIC software program (developed by Yuntao Ji, Institute of Geology, China Earthquake Administration), and the relative displacements parallel to the fault were compared with the thermal variations.

Figure 6 shows the temporal and spatial distributions of the relative displacements of the fault. The range of the analysis region is the same as that of the temperature contour map, but there were a greater number of calculation units. For convenience, the horizontal axis, which indicates the spatial position of the fault, has been converted to be consistent with Fig. 5a. The vertical axis is time, "0" represents the instability time, and negative values represent times prior to the instability. Figure 6 shows three locations of high relative displacement on the left side of the fault that occurred at −4 s. Figure 5a also shows that larger temperature increases occurred at the same time and locations. The local sliding zone expanded from left to right at −2 s (Fig. 6), and Fig. 5a shows the rapid expansion of the warming regions at the same time. Thus, it can be speculated that the increasing temperature on the fault in the quasi-dynamic stage was due to frictional sliding. Because the high relative fault displacement regions in the quasi-static stage did not clearly correspond to the

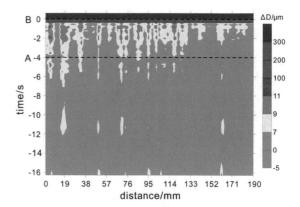

Figure 6
Temporal and spatial distributions of the relative displacement of the fault: The *horizontal axis* represents the distance from the start calculation unit, and the *vertical axis* is time. The *letters* and *dashed lines* mark the divisions of the deformation stages

warming regions and the variation in amplitude was small, it can be inferred that the temperature rise in the quasi-static stage was possibly due to stress concentration.

### 4.2. Common Characteristics of Thermal Variations in the Meta-Instability Stage

The main results discussed in Sect. 3 were obtained by analyzing a single stick slip event "EXP2-SS4". In fact, due to different loading histories, the positions and degrees of stress release in the wall rocks were diverse for the different stick slip events. Figure 7 shows the thermal increment images for the other two stick slip events in "EXP1" and six stick slip events in "EXP2" under the same loading rate but different lateral pressures. The results were obtained using the same data-processing procedures employed for Fig. 3. The top eight images in Fig. 7 were obtained in the quasi-static stage, and the bottom eight images were obtained in the quasi-dynamic stage. The common features of those stick slip events were the large-scale temperature decreases of the wall rocks in the quasi-static stage and the enhancement of temperature increase amplitude and expansion of temperature increase regions on the fault in the quasi-dynamic stage. Due to the different positions and degrees of the pre-slips, the regions and extents of the temperature decreases in the wall rocks in the quasi-static stage differed. In

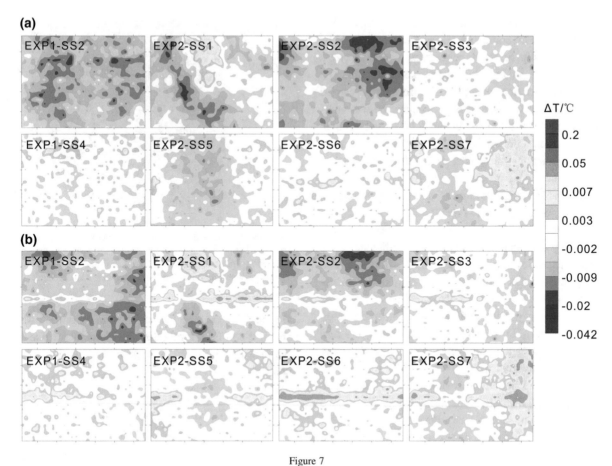

Figure 7

Thermal increment images of the other stick slip events: The *top eight images* were obtained in the quasi-static stage, and the *bottom eight images* were obtained in the quasi-dynamic stage. The temperature increments in each image are relative to time $N$ of each stick slip cycle

addition, due to differences in the pre-slip positions, sliding velocities and normal stresses, the regions and amplitudes of the temperature increases on the fault in the quasi-dynamic stage also differed. In a word, different stick slip events have obvious common thermal field variations at the critical points of different deformation stages, which can be used as thermal indicators to identify the deformation stages of faults.

### 4.3. Possible Applications of Thermal Observations in the Field

Our results have revealed the different thermal field evolutions of a planar strike slip fault and neighboring wall rocks during the meta-instability stage in the laboratory. However, faults in the field are complex and are by no means simple faults with

straight geometries occurring in homogeneous media. The essence of an earthquake is the rapid sliding of a fault, which requires the fault geometry to be relatively straight and the lithology to be uniform in the final stage prior to instability (meta-instability stage). We, therefore, designed a relatively flat fault as the research object to focus on the changes of the final stage prior to instability to avoid the influences of factors such as fault geometry, lithology and the physical environment of the fault zone and to decompose the complex problem into several simple questions and study them separately. We will now only consider the stress changes of a simple fault and discuss the possible applications of thermal observations in the field.

The Kunlun $M_s$ 8.1 earthquake occurred on November 14, 2001, on the eastern Kunlun left lateral strike slip fault. Using remote sensing data,

Chen et al. (2003) found that the brightness temperature inside the fault zone became 1 °C higher than that of the outside region 1.5 months prior to the event and returned to a normal value 2 °C lower than that of the outside region after the event. The result was similar to the phenomenon we observed when entering the quasi-dynamic stage in the laboratory, although when considering heat conduction theory, the causes of temperature increases on faults in the field may involve more complicated mechanisms than friction (Liu et al. 2004b). The $M_s$ 7.0 earthquake hit Lushan County, Sichuan province, on April 20, 2013. Ground temperature measurements in the boreholes near Kangding 95 km from the epicenter showed persistent decreases 3 months prior to the earthquake and abrupt decreases 3 days prior to the earthquake (Chen et al. 2013). Meanwhile, a borehole strain-meter north of Kangding recorded eight tensile strain steps over a range ten times the tidal amplitude a couple of days prior to the earthquake (Chi et al. 2013). These results are similar to the phenomenon of decreasing temperatures in the wall rocks when entering the meta-instability stage. Geological conditions vary in different areas, and the thermal field distributions caused by deformation may be related to earthquake type. Whether such phenomena can be observed prior to other earthquakes needs to be analyzed in combination with specific geological conditions.

## 5. Conclusions

On the basis of thermal full-field observations, this paper has described a more specific short-impending stage in laboratory earthquakes.

1. According to the experimental results, thermal and deformation fields are closely related. Using the locations and amplitudes of thermal variations, the deformation patterns can be estimated. Observations of thermal fields may serve as auxiliary tools for analyses of deformation fields.
2. Pre-slip on a fault can occur at any stage after the deformation curve deviating from linearity, and it is difficult to determine the stress state of a fault based on the occurrence of a pre-slip. From the

theory of meta-instability, the stress state can be identified by the evolution of the thermal field.
3. There are two obvious critical points prior to stick slip instability: the peak stress point and the turning point from the quasi-static stage to the quasi-dynamic stage. The thermal variations of the former occur over a large region of wall rocks with decreasing temperatures and increasing temperatures on sporadic sections of the fault. The thermal variations of the latter are the rapid expansion of temperature increase regions on the fault and the enhancement of temperature increase amplitude. The release of stress in the wall rocks and stress concentrations in some parts of the fault are the principle mechanisms for thermal variations in the quasi-static stage. The spread of slip regions and the acceleration of slip rates are the main mechanisms for thermal variations in the quasi-dynamic stage.
4. In summary, a new method of using the spatial and temporal evolutions of a thermal field has allowed us to gain a better understanding of the physical processes of earthquake preparation and occurrence and may have certain applications for identifying the short-impending stages of earthquakes in the field.

## Acknowledgements

We are grateful to Yanqun Zhuo and Yuntao Ji for their help in performing the experiments. The comments by the anonymous reviewers helped to improve the manuscript. This research is supported by the National Natural Science Foundation of China (Grant No. 41172180,41572181) and Basic Research Funds from the Institute of Geophysics, China Earthquake Administration (Grant No. DQJB15B07).

### REFERENCES

Chen, M. H., Deng, Z. H., & Jia, Q. H. (2003). The relationship between the satellite infrared anomalies before earthquake and the seismogenic fault—a case study on the 2001 Kunlun earthquake. *Seismology and Geology*, 25(1), 100–108. doi:10.3969/j.issn.0253-4967.2003.01.010. **(in Chinese with English abstract)**.

Chen, S. Y., Liu, P. X., Guo, Y. S., Liu, L. Q., & Ma, J. (2015). An experiment on temperature variations in sandstone during biaxial loading. *Physics & Chemistry of the Earth Parts A/b/c, s, 85–86*, 3–8. doi:10.1016/j.pce.2014.10.006.

Chen, S. Y., Liu, L. Q., Liu, P. X., & Ma, J. (2009). Theoretical and experimental study on relationship between stress-strain and temperature variation. *Science China Earth Sciences, 52*(11), 1825–1834.

Chen, S. Y., Liu, P. X., Liu, L. Q., & Ma, J. (2013). A phenomenon of ground temperature change prior to Lushan earthquake observed in Kangding. *Seismology and Geology, 35*(3), 634–640. doi:10.3969/j.issn.0253-4967.2013.03.017. **(in Chinese with English abstract)**.

Chi, S. L., Liu, Q., Chi, Y., Deng, T., Liao, C. W., Yang, G., et al. (2013). Borehole strain anomalies before the 20 April 2013 Lushan $M_s$ 7.0 earthquake. *Acta Seismologia Sinica, 35*(3), 296–303. doi:10.3969/j.issn.0253-3782.2013.03.002. **(in Chinese with English abstract)**.

Cyranoski, D. (2004). Earthquake prediction: a seismic shift in thinking. *Nature, 431*(7012), 1032–1034.

Das, S., & Scholz, C. (1981). Theory of time dependent rupture in the earth. *Journal of Geophysical Research: Solid Earth (1978–2012), 86*(B7), 6039–6051.

Dieterich, J.H. (1986). A model for the nucleation of earthquake slip. Earthquake source mechanics, 37–47.

Dieterich, J. H. (1992). Earthquake nucleation on faults with rate-and state-dependent strength. *Tectonophysics, 211*(1–4), 115–134.

Ellsworth, W., & Beroza, G. (1995). Seismic evidence for an earthquake nucleation phase. *Science, 268*(5212), 851–855. doi:10.1126/science.268.5212.851.

Fang, Z., Dieterich, J. H., & Xu, G. (2010). Effect of initial conditions and loading path on earthquake nucleation. *Journal of Geophysical Research Atmospheres, 115*(B6), 449–463. doi:10.1029/2009JB006558.

Geller, R. J., Jackson, D. D., Kagan, Y. Y., & Mulargia, F. (1997). Enhanced: earthquakes cannot be predicted. *Science, 275*(5306), 1616–1620.

Hasegawa, A., & Yoshida, K. (2015). Preceding seismic activity and slow slip events in the source area of the 2011 $M_w$ 9.0 Tohoku-Oki earthquake: a review. *Geoscience Letters, 2*(1), 1–13. doi:10.1186/s40562-015-0025-0.

Ida, S. (2010). Striations, duration, migration and tidal response in deep tremor. *Nature, 466*, 356–359. doi:10.1038/nature09251.

Jordan, T. H., Chen, Y. T., Gasparini, P., & Madariaga, R. (2011). Operational earthquake forecasting—State of knowledge and guidelines for utilization. *Translated World Seismology, 54*(4), 315–391. doi:10.4401/ag-5350.

Kano, Y., Mori, J., Fujio, R., Ito, H., Yanagidani, T., Nakao, S., et al. (2006). Heat signature on the Chelungpu fault associated with the 1999 Chi-Chi. *Taiwan earthquake. Geophysical Research Letters, 33*, L14306. doi:10.1029/2006GL026733.

Kato, N., & Hirasawa, T. (1996). Effects of strain rate and strength nonuniformity on the slip nucleation process: a numerical experiment. *Tectonophysics, 265*(3), 299–311.

Kato, N., Yamamoto, K., Yamamoto, H., & Hirasawa, T. (1992). Strain-rate effect on frictional strength and the slip nucleation process. *Tectonophysics, 211*(1–4), 269–282.

Liu, L. Q., Chen, G. Q., Liu, P. X., et al. (2004a). Infrared measurement system for rock deformation experiment. *Seismology and Geology, 26*(3), 492–501. doi:10.3969/j.issn.0253-4967.2004.03.013. **(in Chinese with English abstract)**.

Liu, P. X., Liu, L. Q., Chen, S. Y., et al. (2004b). An experiment on the infrared radiation of surficial rocks during deformation. *Seismology and Geology, 26*(3), 502–511. doi:10.3969/j.issn.0253-4967.2004.03.014. **(in Chinese with English abstract)**.

Ma, J., Guo, Y., & Sherman, S. I. (2014). Accelerated synergism along a fault: a possible indicator for an impending major earthquake. *Geodynamics Tectonophys, 5*(2), 387–399. doi:10.5800/GT2014520134.

Ma, J., Liu, L. Q., Liu, P. X., & Ma, S. L. (2007). Thermal precursory pattern of fault unstable slip: an experimental study of en echelon faults. *Chinese Journal of Geophysics, 50*(4), 1141–1149.

Ma, S. L., Liu, L. Q., Ma, J., Wang, K. Y., Hu, X. Y., Liu, T. C., et al. (2003). Experimental study on nucleation process of stick-slip instability on homogeneous and non-homogeneous faults. *Science China Earth Sciences, 46*(2), 56–66.

Ma, J., Ma, S. P., Liu, L. Q., & Liu, P. X. (2010). Experimental study of thermal and strain fields during de-formation of en enchelon faults and its geological implications. *Geodynamics & Tectonophysics, 1*(1), 24–35.

Ma, J., Sherman, S. I., & Guo, Y. S. (2012). Identification of meta-instable stress state based on experimental study of evolution of the temperature field during stick-slip instability on a 5° bending fault. *Science China Earth Sciences, 55*(6), 869–881. doi:10.1007/s11430-012-4423-2.

McLaskey, G. C., & Lockner, D. A. (2014). Preslip and cascade processes initiating laboratory stick slip. *Journal of Geophysical Research: Solid Earth, 119*(8), 6323–6336. doi:10.1002/2014JB011220.

Mei, S. R. (1986). The precursory complexity and regularity of the tangshan earthquake. *Earth, Planets and Space, 34*(Supplement), S193–S212.

Melgar, D., Fan, W. Y., Riquelme, S., Geng, J. H., Liang, C. R., Fuentes, M., et al. (2016). Slip segmentation and slow rupture to the trench during the 2015, $M_w$ 8.3 Illapel, Chile earthquake. *Geophysical Research Letters, 43*, 961–966. doi:10.1002/2015GL067369.

Ohnaka, M., Kuwahara, Y., & Yamamoto, K. (1987). Constitutive relations between dynamic physical parameters near a tip of the propagating slip zone during stick-slip shear failure. *Tectonophysics, 144*(1), 109–125.

Okubo, P. G., & Dieterich, J. H. (1984). Effects of physical fault properties on frictional instabilities produced on simulated faults. *Journal of Geophysical Research:solid Earth, 89*(B7), 5817–5827. doi:10.1029/JB089iB07p05817.

Rogers, G., & Dragert, H. (2003). Episodic tremor and slip on the Cascadia subduction zone: the chatter of silent slip. *Science, 300*(5627), 1942–1943. doi:10.1126/science.1084783.

Thompson, B. D., Young, R. P., & Lockner, D. A. (2009). Premonitory acoustic emissions and stick-slip in natural and smooth-faulted Westerly granite. *Journal of Geophysical Research Atmospheres, 114*(2), 1205–1222. doi:10.1029/2008JB005753.

Wu, L. X., Liu, S. J., Wu, Y. H., & Wang, C. Y. (2006). Precursors for rock fracturing and failure—Part I: IRR image abnormalities. *International Journal of Rock Mechanics and Mining Sciences, 43*(3), 483–493. doi:10.1016/j.ijrmms.2005.09.002.

Wu, K. T., Yue, M. S., & Wu, H. Y. (1976). Certain characteristics of Haicheng earthquake ($M = 7.3$) sequence. *Chinese Journal of Sinica, 63*(8), 265–267.

Yasuo, Y., Masao, N., Makoto, N., Joachim, P., Christoph, J., Takayoshi, W., et al. (2015). Nucleation process of an M2 earthquake in a deep gold mine in South Africa inferred from on-fault foreshock activity. *Journal of Geophysical Research Solid Earth, 120,* 5574–5594. doi:10.1002/2014JB011680.

Yin, X. C., Chen, X. Z., Song, Z. P., & Yin, C. (1995). A new approach to earthquake prediction: the load/unload response ratio (LURR) theory. *Pure and Applied Geophysics, 145*(3), 701–715.

Yin, X. C., Wang, Y. C., Peng, K. Y., Bai, Y. L., Wang, H. T., & Yin, X. F. (2000). Development of a new approach to earthquake prediction: load/unload response ratio (LURR) Theory. *Pure and Applied Geophysics, 157,* 2365–2383.

Yin, X. C., Zhang, L. P., Zhang, Y., & Peng, K. (2008). The newest developments of load-unload response ratio (LURR). *Pure and*

*Applied Geophysics, 165*(3), 711–722. doi:10.1007/s00024-008-0314-z.

Zhang, G. M., & Fu, Z. X. (1990). Discussions on complexity of earthquake precursors from a point of rock instability. *Journal of Seismological Research, 13*(3), 215–222. **(in Chinese with English abstract)**.

Zhuo, Y. Q., Guo, Y. S., Yuntao, J. I., & Jin, M. A. (2013). Slip synergism of planar strike-slip fault during meta-instable state: experimental research based on digital image correlation analysis. *Science China Earth Sciences, 56*(11), 1881–1887.

(Received  February 12, 2017, revised  July 2, 2017, accepted  July 19, 2017)

Pure Appl. Geophys.
© 2018 Springer International Publishing AG, part of Springer Nature
https://doi.org/10.1007/s00024-018-1885-y

**❚ Pure and Applied Geophysics**

CrossMark

# Hydrogeological and Geochemical Observations for Earthquake Prediction Research in China: A Brief Overview

Bo Wang,[1] Yaowei Liu,[2] Xiaolong Sun,[2] Yuchuan Ma,[1] Lei Zhang,[2] Hongwei Ren,[2] and Zhen Fang[3]

*Abstract*—Hydrological and geochemical changes have been studied extensively in China for the purpose of earthquake prediction since 1966. This paper gives a brief overview of the findings, which have been reported mainly in Chinese so far. Parameters studied include water level; water temperature; and radon, hydrogen and mercury concentrations. Significant changes have been observed before some large or even intermediate-sized earthquakes, among many cases that did not show any such changes. Background noise caused by other environmental variables and possible mechanisms for the pre-earthquake anomalies are discussed. Due to the complexity of earthquake prediction, much more monitoring efforts as well as sophisticated analytic and synthetic efforts are needed in the future.

**Key words:** Earthquake prediction, water level, water temperature, radon, hydrogen.

## 1. Introduction

Scientists have found that hydrogeological and geochemical changes can be observed in wells or springs before earthquakes (Hauksson 1981; King 1980, 1984; King et al. 2006; Freund 1985; Roeloffs 1988; Kissin and Grinevsky 1990; Toutain and Baubron 1999; Cicerone et al. 2009; Ingebritsen and Manga 2014). China has conducted numerous researches on possible precursory anomalies and earthquake prediction using hydrogeological and geochemical observations for 50 years.

China is one of the most seismically active countries in the world. Destructive earthquakes pose a major threat to lives and properties in almost all of the Chinese territory (Huang et al. 2017). In addition to the exploration of scientific problems and accumulating research data, China also has to carry out earthquake prediction, which is significantly different from other countries. China has been conducting systematic research on the relationship between earthquakes and groundwater after the Xingtai $M_s$ 6.8 and $M_s$ 7.2 earthquakes of Hebei Province in 1966. According to the *Earthquake Cases in China* from 1966 to 2012, there were about 1162 underground fluid anomalies before 170 earthquakes (Sun et al. 2016a, b).

China has built a systematic monitoring network of groundwater for earthquake prediction in the past 50 years. It contains hydrogeochemical observation stations (Rn, $H_2$, He, Hg, ion composition, etc.), groundwater observation stations (water level, flux) and geothermal observation stations (water temperature, ground temperature). Up to now, more than 684 observation stations have been set up in almost all provinces of China, with a total of 1522 observation items, including 486 water level observations, 465 water temperature observations, 219 radon observations, 15 hydrogen observations of faults, 73 mercury observations and 264 other observations (Fig. 1).

The purpose of this paper is to illustrate the research process of the relationship between groundwater and earthquakes in China in the past decades, summarize the successful experience and inadequacy in the study, and list some aspects that should be studied in the future. It is also hoped that international experts in this field would provide good suggestions on the research of earthquake prediction in China.

---
[1] China Earthquake Networks Center, Beijing 100045, China.
[2] Key Laboratory of Crustal Dynamics, Institute of Crustal Dynamics, China Earthquake Administration, Beijing 100085, China. E-mail: liuyw20080512@126.com
[3] Anhui Earthquake Administration, Hefei 230031, China.

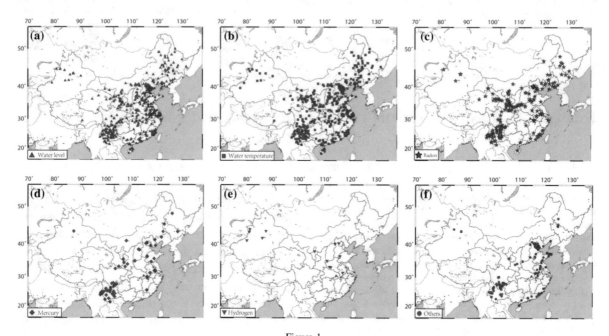

Figure 1
The distribution of subsurface fluid observation stations for earthquake prediction. **a** Water level, **b** water temperature, **c** radon, **d** mercury, **e** hydrogen, **f** others

## 2. Hydrogeological Observations

The analysis of the hydrogeological changes in wells or springs is an important method in earthquake prediction research, which aims at capturing crustal strain information before the occurrence of earthquakes. The correlation between the change in groundwater and crustal strain has been a focus of scientific research in recent years.

### 2.1. Water Level

Water level observation is superior in number to other items in the underground fluid monitoring network center of China. Water level observation instruments have gone through many stages, and now LN-3(A) digital instrument, designed by the Institute of Earthquake Science of China Earthquake Administration with a resolution of 1 mm, has been widely used for water level measurement in the affiliated departments of China Earthquake Administration.

The response of water level to the atmospheric pressure is prevalent: 54% of the wells have obvious air pressure effect (Che et al. 2006). The tidal

phenomena are also observed in groundwater wells in China (Cai 1980; Li et al. 1990; Zhu et al. 1997). To remove the environmental variables that influence underground water level observations, scientists have used a variety of different data processing methods, such as regression analysis, wavelet analysis method, spectrum analysis and harmonic analysis, which have been widely used in the analysis of water level data (Tian and Gu 1985; Zhang et al. 1986a, b, 1989b, c, d; Che et al. 1990; Yu et al. 1990; Du 1991; Zhang et al. 1997; Yan et al. 2007; Yan 2008; Lai 2014). In addition, some scholars have found that the earth tide characteristics of water level in wells always changed before earthquakes. For example, Shi and Wang (2013) calculated the tidal factors and phase lags of four selected wells and found that the tidal factors of the four wells were in a rapidly changing stage before the Wenchuan 8.0 earthquake in 2008, but the phase lag and the tidal residual phase were in a rapidly declining stage, which is a good reflection of the stress in the well aquifer.

Classification and analysis on the water level changes caused by big earthquakes have been carried out in recent years (Yan and Huang 2009). A great deal of research has also been made on the variation of permeability coefficient of aquifer caused by seismic waves and the role of dynamic stress and static stress in the change of groundwater level (Liu et al. 2011; Lai et al. 2011; Shi et al. 2013; Shi and Wang 2015). Besides, some researchers have studied the influence on future seismicity, by analyzing the changes in groundwater caused by big earthquakes and claim that the concentration area of coseismic response of water level is always the area where earthquakes are prone to occur in the future (Li 1995; Huang et al. 2000; Wang 2000).

The sharp rise or fall in water level to the exclusion of other environmental variables is generally considered to be a possible short-term precursory anomaly of earthquake. It may appear as a frequent step-like change; for example, Cao and Zhang (1999) analyzed and proved that the frequent step-like changes in Wanquan well have a good correlation with the nearby regional seismic activities (Fig. 2); it may also appear as a rising trend. For example, the water level of Fengzhen well suddenly rose 40 mm, 62 days before the Zhangbei $M_s$ 6.2 earthquake on January 10, 1998, and then continued to rise by 60 mm until the earthquake occurred (Ding and Ha 2012). In addition to the short-term anomalies, water level observations also show a monthly variation and

annual variation. There are also a lot of researches on the spatial and temporal characteristics between the abnormal changes of groundwater level and seismic activities (Liu et al. 1999; Cao et al. 2002; Yang et al. 2009).

Due to the block structure of China mainland, researches have been carried out on the changes of groundwater level related to typical blocks, seismic zones or large earthquakes. For example, Liu et al. (2002) analyzed the short-term anomalies of groundwater in the Sichuan–Yunnan region and summed up the temporal characteristics and spatial distribution of the precursory phenomena. Wang (2004) analyzed the dynamic anomalies of the groundwater in the northern Tianshan Mountain before the six $M > 7$ earthquakes in Xinjiang and neighbor areas since 1980 and found the short-term pre-earthquake changes of groundwater level before three earthquakes, with a prominent coseismic and post-seismic responses. Before the Ludian $M$ 6.5 earthquake of 2014, the long-term and medium-term anomalies of groundwater were mainly in the range of 300–500 km from the epicenter, while the short-term anomalies were concentrated within 100 km from the epicenter (Liu et al. 2015).

Groundwater dynamics is closely related to the geological environment and tectonic background and

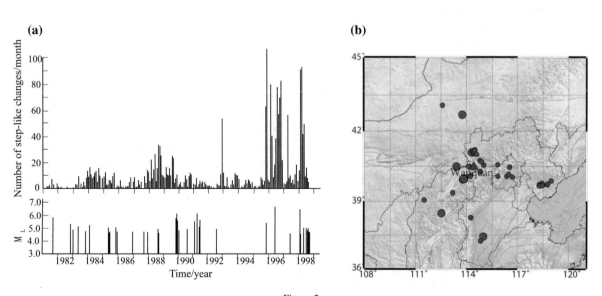

Figure 2
**a** Number of step-like changes of groundwater level in Wanquan well and its relationship with regional seismic activities [after Cao and Zhang (1999)]; **b** the location of earthquake epicenter and Wanquan station

also affected by meteorological factors and human activities. In the study of precursory observation, the specific causes of water level changes in observation wells should be analyzed in combination with the structural location and geographical environment of the wells, so as to determine whether they are precursory anomalies associated with crustal activities.

## 2.2. Water Temperature

The water temperature precursory research with high-precision instruments in China was launched in 1979. The quintessence of this research is to explore the crustal geothermal change process caused by the migration of hot material, the transformation of media energy and the change of hydrodynamic conditions under the action of tectonic stress, by monitoring and analyzing water temperature observed in different wells and springs. In springs, the water temperature was measured by an observer, usually with an alcohol thermometer, once a day at a similar time, while the continuous monitoring of groundwater temperature in wells was recorded by a quartz thermometer developed by the Institute of Crustal Dynamics, China Earthquake Administration. The sampling rate is 1 min, and precision is 0.05 with a resolution up to 0.0001 °C.

The dynamic characteristics of groundwater temperature are affected by the surrounding environmental conditions. Therefore, the characteristics of temporal variations in groundwater temperature differ from well to well and can be divided into different types based on different frequency bands. Many scientists have made statistical analysis on the morphological characters of groundwater temperature in different observation wells (Yu et al. 1997; Che et al. 2003, 2008; Che and Yu 2006; Zhang et al. 2007; Zhao et al. 2009; Ma 2015). As mentioned in the previous section, water level in a confined well shows a good response to earth tide, and such similar fluctuations of water temperature were also observed in many confined wells. It indicates that the observation of groundwater temperature is also affected by solid tide. In addition, there are many meteorological factors and human factors influencing the daily variations of groundwater temperature, such as rainfall, groundwater exploitation and instrument calibration. So, the extraction of such factors is the basis for a better understanding of the earthquake precursor using groundwater temperature.

Due to the low thermal conductivity of rocks, it takes a long time to reach the earth's surface for the deep water temperature changes in the crust. Therefore, the mechanism of heat conduction cannot explain the wide variation of water temperature. However, the heat convection of underground fluid can effectively change the transfer for deep temperature, especially in areas with active faults. Due to the enhancement of tectonic activity, it may increase or decrease the hydrodynamic characteristics of active faults, resulting in groundwater temperature rise or fall (Yu et al. 1997; Gottardi et al. 2013). From this viewpoint, the possibility of groundwater temperature anomalies is higher in the geothermal zone or in the tectonic zone.

According to Wan et al. (1993), the water temperature of Rewu spring is stable at 20–22 °C under normal conditions. After May 20, 1982, the water temperature rose day by day and up to 30 °C in early June. On June 16, an earthquake of $M$ 6.0 occurred in Ganzi, Sichuan Province, 310 km away from the observation point. After the earthquake, the water temperature continued to rise, up to 38 °C at the end of June, and an earthquake of $M$ 5.3 struck there again. Similarly, before the Datong $M$ 6.1 earthquake on October 19, 1989, the water temperature of Sanmafang well, 60 km away from the epicenter, showed an upward change, which was 25 times that of normal fluctuations. After the earthquake, the water temperature rose slowly and returned to normal about 20 days later (Chen et al. 1994). We observed similar pre-earthquake changes of the water temperature in Yushu well, China, before the Wenchuan $M$ 8.0 earthquake in 2008 and Yushu $M$ 7.0 earthquake in 2010 (Fig. 3). Furthermore, the changes of water temperature in Yushu well have a good correlation with the earthquakes that occurred in the Tibetan block and its margins (Wang et al. 2016; Sun et al. 2017). The China government launched the Wenchuan Fault Scientific Drilling Project after the Wenchuan $M$ 8.0 earthquake in 2008. The results showed that continuous long-term

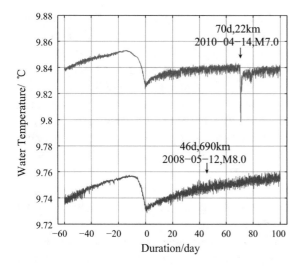

Figure 3
The changes in the groundwater temperature in Yushu well before the Wenchuan *M* 8.0 earthquake and the Yushu *M* 7.0 earthquake

changes of water temperature in drills had a better link to the low speed friction (Li et al. 2015).

In addition to the study of abnormal changes before a single observation well, some scholars have studied the regional water temperature changes in multiple observation wells. Li et al. (2017) analyzed the data of water temperature and $M \geq 5.0$ earthquakes in Yunnan province using the subordinate function analysis method and the R-value evaluation method. It was considered that the regional water temperature changes can be better used for earthquake prediction. Zhang et al. (2014a, b) analyzed the changes of water temperature before 49 earthquakes in Yunnan province, classified them into four types and concluded that the water temperature anomalies always occurred 50–170 days before the earthquake. The smaller the magnitude, the smaller is the time interval between the abnormal time and the occurrence of the earthquake. However, when the magnitude of the earthquake is large, there is no greater distribution of time intervals.

There are usually three kinds of water temperature precursors in the length of duration ($t$) before earthquakes: imminent anomaly ($t \sim 1$ month); short-term anomaly ($t \sim 3$ months); medium-term anomaly ($t \sim 1$ year or more) (Liu et al. 2008). However, the three anomalies do not have specific morphological changes. It can show a rise change, i.e., the value of water temperature continues to rise from days to months before the earthquake, and the amplitude is several times to ten times that of the daily variation; It may also show a drop change, i.e., the value of water temperature declines before the earthquake, and its amplitude is more than several times the daily variation. Sometimes, it can also appear as a sudden rise or fall and return to its original value in hours or days, with an abnormal amplitude of several to ten times that of the average daily variation. Yan et al. (2015) studied the abnormal changes of five hot springs in Western Sichuan before the Lushan *M* 7.0 earthquake and concluded that the Litang hot spring and Daofu hot spring, far from the epicenter, showed medium-term anomalies 2 years before the earthquake; two hot springs in Kangding area, close to the epicenter, showed short-term anomalies 3 months before the earthquake. This indicted that groundwater temperature anomalies gradually converged near the epicenter before the earthquake and may be of some significance for the location of earthquake prediction.

With the study on thermal infrared precursor, as well as exploring and monitoring heat anomalies in the active fault zone (Ma et al. 2015a, b; Xie et al. 2015; Zhang et al. 2017), further advances have been made in the study of earthquake prediction using water temperature in recent years.

## 3. Geochemical Observations

Research on fault gas was initiated in the 1980s and attracted widespread scientific interests. With the development of observation techniques and theoretical knowledge of geochemistry, geochemical observation of faults has become a hotspot once more in recent years. Radon, mercury, hydrogen, methane, carbon dioxide, etc., are used for geochemical observations (Cui et al. 2012; Fang et al. 2012; Che et al. 2015; Zhang et al. 2015; Guo et al. 2016; Sun et al. 2016a, b).

### 3.1. Radon

Radon is the decay daughter nuclide of radioactive uranium, radium and thorium in the crust. There exist three radioactive isotopes in nature: $^{219}$Rn, $^{220}$Rn, $^{222}$Rn. $^{222}$Rn has a half-life of 3.825 days, while the other two have a half-life of 3.96 and 55.65 s, respectively. As the two isotopes of short half-life are not suitable for earthquake research, the one we studied was $^{222}$Rn.

Radon migration is closely related to the external environment. The active fault zone, because of the large porosity and high permeability, becomes a good migration and concentration channel of fault gas. Since the discovery of radon concentration that escaped from the upper part of the fault has been linked to seismic activity in the 1970s, many scientists have begun to observe changes in radon concentrations to predict earthquake (King 1980; Wang and Li 1991; Shi and Zhang 1993; Zhang and King 1994; Zmazek et al. 2003; Singh et al. 2010).

In 1968, the first geochemical observation station in China was established in Xinji, Hebei province (Zhang et al. 1988). Later, a comprehensive observation station was established in Xiaotangshan, Beijing, especially for the observation of dissolved radon in hot water (Cai 1980). In the middle 1990s, the successful development of the digital SD-3A automatic radon measurement instrument accelerated the study of radon and until now the SD-3A is still most popular for continuous observation of radon by scientists in the earthquake field in China.

Long-term observations show that radon concentrations can change significantly before earthquakes. It may appear to decline before an earthquake: for example, about 65 days before the $M$ 6.8 earthquake occurred near the coast of Taiwan's Taitung, the concentration of water radon from the observation station 20 km away began to decline (Kuo et al. 2006); There have been such similar declining changes of radon concentration in Jiayuguan station before earthquakes nearby (Zhang et al. 2004; Wang et al. 2011). It has also been observed that there are persistent high anomalies of radon concentration before earthquakes: for example, several days before the Gonghe $M$ 7.0 earthquake in 1990, the concentrations of radon in the stations 250 km away from the epicenter were characterized by high anomalies (Shi and Wang 2000). Due to the massive observation data of radon, some comparative studies have also been made on the changes of radon concentration before and after earthquakes (Liu and Ren 2009; Singh et al. 2010).

Some studies have shown that changes of radon concentration actually reflect the variation of crustal stress. The aquifer is a porous medium, which deforms at a very low stress level, thus changing the porosity and pore pressure, resulting in changes in the hydrodynamic state and thereby causing radon migration (Shi and Zhang 1993; Che et al. 1997). On one hand, radon escapes from the crust by the effect of diffusion, suction or convection. In addition, Wu et al. (1997) considers that diffusion and convection can only account for short-distance migration of radon, while long-distance migration mainly depends on the action of relay transmission. On the other hand, under the action of horizontal and vertical motion of groundwater, radon reaches the surface along the fracture or fissure (Li and Xu 1990).

In summary, changes of radon concentrations are affected by a variety of factors and dominated by a combination of multiple migration mechanisms. Radon has different short-term performances before earthquakes. The study of the influencing factors of radon concentration and the discussion on the migration mechanism of radon can better our understanding of its role in earthquake prediction research.

### 3.2. Hydrogen

Hydrogen has characteristics of being lightweight, slight solubility, extremely easy mobility, etc. The content of hydrogen in the atmosphere is very low, with a stable concentration of 0.5 ppm. Previous studies suggested that there may be a great amount of hydrogen which may have been sealed inside the earth when it was initially formed and released from the active fault (Gold and Soter 1980; Du et al. 1995; Chen 1996).

In 1980, Wakita et al. (1980) found the correlation between the changes in hydrogen concentration and fault activities. Thereafter, scientists in USA observed that hydrogen escaped from the San Andreas Fault and found that it was associated with microseismic activity (McGee et al. 1982, 1983; Ware et al. 1984; Sato et al. 1986). Since then, more and more studies on fault hydrogen observation for earthquake monitoring have been made in some countries (Sato and McGee 1981; Satake et al. 1984;

Sato et al. 1984; Sugisaki and Sugiura 1985; Wang and Li 1991; Dogan et al. 2007).

China began to observe the dissolved hydrogen in the 1970s. The dissolved hydrogen in a hot water well of Beijing was higher before the Ninghe $M$ 6.9 earthquake on November 15, 1976, and decreased gradually after the earthquake (Wang et al. 1982). Shangguan (1989) measured the components of fault gases in western Yunnan, four of which had high hydrogen concentration, and all of the four points were at the intersection or in the vicinity of active faults, where there have been several $M \geq 6$ strong earthquakes. As hydrogen is mainly derived from the deep crust, long-term observations show that in the aseismic period, the concentration of hydrogen fluctuates only in a very small range. For example, several moderate earthquakes occurred in the west of Gansu province from 1987 to 1988, and the concentrations of many fault gases, including hydrogen, were significantly increased before these earthquakes (He et al. 1990). Gao and Fan (1992) conducted gas measurement on the Xiadian Fault; the hydrogen concentration in the fault was higher several days before the Tongxian $M$ 2.6 earthquake and Haituo Mountain $M$ 4.8 earthquake in 1990, respectively. Wang et al. (1994) have collected data of hot spring gas in the Lancang and Batang earthquake regions and found $H_2$ was sensitive to seismic activity. The concentration decreased obviously with the increase of distance from the epicenter, indicating that the stress effect resulting from earthquakes decreased with the increase of distance from the epicenter to produce less released gas.

After the Wenchuan $M$ 8.0 earthquake in 2008, China carried out a scientific deep drilling exploration project. Scientists found that the abrupt high value of hydrogen concentration was related to the subsequent earthquakes: for example, 20 days before the Yushu $M$ 7.1 earthquake on April 14, 2010, the hydrogen concentration began to rise and continued for 7 days in the WFSD-2 well (with a epicentral distance of 680 km), which was drilled to 642.36–676.22 m. The Yushu $M$ 7.1 earthquake occurred 13 days after the measured value and returned to the normal background (Li et al. 2015). Fan et al. (2012) found that the hydrogen concentration in the Xiaxian Fault, 330 km away from the

epicenter, had a value of more than 20 times the normal background value 7 days before the Taikang $M$ 4.6 in 2010 and $M$ 4.1 earthquake in 2011, respectively (Fig. 4).

Non-continuous field measurements of fault gas were also conducted. Zhou et al. (2010) conducted a fault gas measurement near the Wenchuan $M$ 8.0 earthquake fault zone, including $H_2$, He, $CO_2$, $CH_4$, $O_2$, $N_2$, Rn, Hg, etc. The maximum concentrations of hydrogen and helium measured in June 2008 were in the region with large vertical dislocation, and the concentration of hydrogen and helium decreased gradually with the aftershock's decrease. It was demonstrated that the release of hydrogen in fault gas has a certain genetic relationship with fault activity and seismic activity.

Although the background value of hydrogen concentration is different for each fault or region, the concentration of hydrogen tends to rise sharply by several times that of the background value before earthquakes (Li et al. 2009; Zhou et al. 2011). Anomalies of hydrogen usually occurred a few months before earthquakes, but there is no accurate correspondence between the variation of hydrogen concentration and the magnitude of earthquake. On one hand, it shows that hydrogen is really a reflection of faulting and regional stress; on the other hand, it shows the complexity of hydrogen as an earthquake precursor.

## 3.3. Mercury

Mercury has a bouncy chemical property with a strong redox, existing in three kinds of valence: $Hg^0$, $Hg^+$ and $Hg^{2+}$. Mercury plays an important role in revealing the relationship between fluid in the fault zone and the occurrence of earthquakes. The mercury in the lower crust and the upper mantle migrate to the surface along the fracture or rock fissure by the pore fluid driven by the hydraulic gradient and forms the abnormal anomalies in rock, soil and groundwater above the fault zone. Therefore, the observation of mercury concentration has been widely used for earthquake prediction and active fault detection (Meng et al. 1997; Wang et al. 2004; Zhang et al. 2014a, b).

The gaseous mercury is easily adsorbed in the fault mud during the ascending process, so the

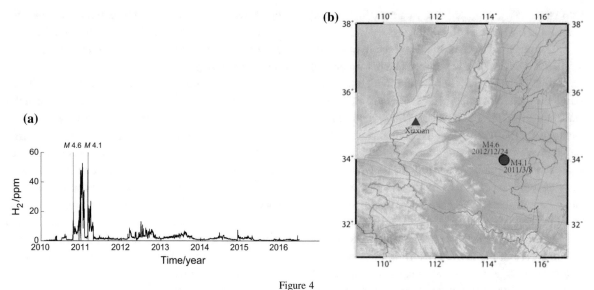

Figure 4
**a** Variations of fault hydrogen concentration in Xiaxian station; **b** the location of earthquake epicenter and Xiaxian station. The red lines represent the Taikang *M* 4.6 earthquake on October 24, 2010, and Taikang *M* 4.1 earthquake on March 8, 2011, respectively

mercury concentration in the fault zone is used to determine the main sliding surface and the rupture process of earthquake. Scientists found that there was a high concentration of mercury in the core samples of the main rupture of the Wenchuan *M* 8.0 earthquake. The content of mercury is mainly related to the degree of rock fracture and the clay mineral inside. The distribution of mercury concentration indicates that deep fluid may have played a role in the occurrence of the Wenchuan *M* 8.0 earthquake (Xu et al. 2008; Li et al. 2013).

The extent of fragmentation and the opening and closure of the fracture can cause uneven distribution of mercury, and there usually is a higher mercury concentration in the hanging wall of the fault (Cheng 1997). The mercury concentrations in carbonaceous siliceous rocks are always higher ($> 100$ ng g$^{-1}$), which are related to the rising and adsorption of mercury along the fracture (Chi 2004).

For the factors affecting gaseous mercury concentration, Li et al. (2009) studied the soil gas in the Yanhuai basin of North China and found that the distribution of mercury was affected by factors such as regional tectonics, stratigraphy and gas exchange rate. The mercury in the soil is also affected by temperature. The vaporing content of mercury in the soil is reduced as a result of the enhanced adsorption

of solids in winter, and when the soil temperature rises in spring, the absorbed mercury is released (Klusman and Jaacks 1987). Also, soil moisture, vegetation and precipitation have some influence on the concentration of mercury (Tao et al. 1992).

Mercury is easily enriched in water, so it is an indicator of seismicity that can be observed from the mercury concentration changes in the wells (springs) in the fault zone. The concentration of mercury in groundwater is affected by a variety of factors, such as the lithology of the surrounding rock, the redox of water, the temperature, the total organic content, the number of suspended particulates and the conductivity of the aquifer (Bagnato et al. 2009; Johannesson and Neumann 2013).

Annual fluctuation is the most common characteristic of mercury concentration, with a higher value in summer and a lower one in winter (Che et al. 2006). Like hydrogen, the concentration of mercury always has a higher value before an earthquake, which is always several times that of the background value. For example, before the Ninglang *M* 5.4 earthquake of Yunnan province in 1988, the mercury concentrations in the Yanyuan well and the Xichang well, about 200 km from the epicenter, began to rise to several times that of the background value, which continued for nearly 100 days (Zhang et al. 1989a)

(Fig. 5). According to the abnormal duration of mercury concentration and its range of distribution, it is possible to estimate the magnitude and onset time of the coming earthquake. Zhang et al. (1989a) suggested that the anomalous range of mercury increases with the magnitude of the earthquake, and the anomalous range for the $M \leq 5.0$ earthquake is limited to 200 km. Therefore, mercury observation can play a significant role in the earthquake short-term monitoring and forecasting.

There are still many uncertain factors that affect the mercury concentration, so it is difficult to scientifically determine whether the mercury concentration anomalies are due to changes of the surface environment, or the migration process of deep seated fluid. Therefore, there is an urgent need for new means to eliminate the uncertainties in observations and further researches on the changes of mercury concentration and seismic activities.

### 3.4. Others

The other observation items for earthquake monitoring are helium, carbon dioxide, methane, etc.

Scientists have made a reasonable prediction of the Zhangbei $M$ 6.2 earthquake based on the typical anomaly of $CO_2$ in the Huailai fault and other subsurface fluid observations nearby (Lin et al. 1998). Some argue that changes in soil gas and thoron concentrations are the most effective tools for seismic monitoring (Yang et al. 2005).

The study shows that the presence of fault accelerates the flow of fluid, leading to a higher concentration of fault gases. Study of the source and migration mechanism of these gases may help to accurately determine the cause of the dynamic changes of underground fluid and provide a more scientific basis for earthquake prediction.

### 4. Obstacles and Challenges

In the past 50 years, Chinese seismologists have gained some experience in the practice of earthquake monitoring and prediction. Using these empirical methods, Chinese seismologists had successfully predicted the Haicheng $M$ 7.3 earthquake in 1975, which greatly reduced casualties and property losses

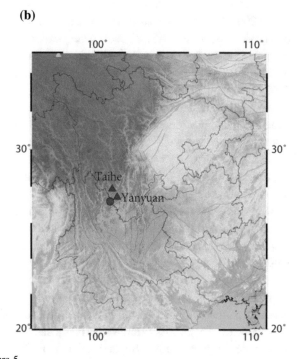

Figure 5

**a** The concentrations of mercury in Yanyuan well (A) and Taihe well (B) before the Ninglang earthquake [after Zhang et al. (1989a)]; **b** the location of the earthquake epicenter and two monitoring sites

(Mogi 1985). However, Chinese seismologists have failed to predict the Tangshan $M$ 7.8 earthquake of 1976 and the Wenchuan $M$ 8.0 earthquake of 2008. More and more seismologists and the public have become aware of the difficulties and challenges involved in earthquake prediction research, especially for short-term earthquake prediction (Mogi et al. 2000).

The international seismologist's discussions of the "silver bullet"—diagnostic precursors for earthquake prediction—involve a wide range of controversies (Felzer and Brodsky 2006). In late 1980s, the earthquake prediction Committee of International Association of Seismology and Physics of the Earth's Interior (IASPEI) established a detection program to make an assessment of the precursors. Only 5 items were included in the initial list of precursors among the 40 listed precursors, and in the final summary in 1997, the chairman of the committee proclaimed none of the precursors had been clearly understood that could be used to predict earthquakes and they were only a collection of phenomena with more than average opportunities for earthquake prediction (Field et al. 2003). The debate on the precursors of earthquakes became more intense in the late 1990s (Field 2007). Some critics concluded that none of the precursors issued to IASPEI and other scientific journals had proved to be a significant diagnostic precursor (Fedotov 1968; Fleischer 1981; Field et al. 2009; Campbell 2009; Cicerone et al. 2009).

A number of research projects for earthquake monitoring and prediction have been implemented in China in recent years (Mileti and Darlington 1997), exploring a variety of observational items and techniques. In addition to groundwater and geochemical observation, there are also other observations at stations, such as crustal deformation, stress and strain, gravity, geomagnetism and geoelectricity. Although there are many precursory anomalies observed before moderate earthquakes in China (Harris et al. 1996), there still exist intractable problems unsolved. For example, there is no established causal correlation between parameters such as the epicenter distance, the time when the precursory anomalies begin and the magnitude in most cases. In case of radon, the porosity and permeability of rock is an important factor affecting radon accumulation in groundwater

and soil (Hamada 1993), so the geological heterogeneity could lead to temporal and spatial variation of radon, and the correlation between precursors and earthquakes remains to be determined (Hamilton 1997; Han and Kamber 2006). These anomalies are merely recorded by one or two instruments, and they are often recorded at monitoring stations far away. Such changes were not recorded in the stations near the epicenter. There are few systematic studies on false and missing alarm rate reports. Therefore, earthquake prediction research in China is empirical and exploratory. Only if we can predict an impending earthquake (location, time and magnitude) with very high probability, the changes could be considered to be a diagnostic precursor, which are prerequisite and helpful for a successful earthquake prediction.

## 5. Conclusions

China has undertaken numerous efforts in earthquake prediction research, although it is a worldwide tricky puzzle. As mentioned in the text, we have made some achievements in earthquake prediction using hydrogeological and hydrogeochemical observations, but have yet encountered many stubborn problems.

As for the observation of water level and water temperature, the biggest problem is the dynamic characteristics of groundwater, which is greatly affected by environmental factors. The groundwater balance feature observed is often not consistent with the hydrodynamic model, bringing difficulties for quantitative analysis of water level precursors caused by earthquakes. Even if the observation wells are not affected by exploitation, as the groundwater runoff area crosses a number of fault zones, the change in the water conduction characteristics of the fault zone directly affect the water level performance of the wells. Although the probability of correspondence between anomaly and earthquake is higher, it is still difficult to give a reasonable explanation for the mechanism. The difficulty we encountered in the study of geochemical observations is the combination of hydrogeochemical model with the mechanical and stress–strain models of tectonic activity.

To advance the study of precursor using underground fluid observations, we should proceed from the following aspects:

1. More water level observation of confined aquifer based on the stress–strain characteristics of rock stratum should be made, and the coupling relation between hydrodynamic model and stress–strain model should be studied. Researches on observations of water level in the same hydrogeological unit combined with the short-term stress–strain characteristics of the regional structure should be augmented.

2. We could do researches on the relationship between subsurface thermal activity and seismogenic process based on the mechanism of aquifer hydrothermal dynamics.

3. Based on the development of hydrogeochemistry and gas geochemistry, the study on the characteristics of $H_2$, Rn, Hg, $CO_2$, microelement, tracer isotope element and other gas components during the development of rock fissure and even rock fracture should be enhanced, to reveal the possible precursory anomalies in the short-term stage of seismogenic process.

## Acknowledgements

We would like to thank Professor Chi-Yu King and three other reviewers for their valuable comments for improving the manuscript. This study was financially supported by the National Natural Science Foundation of China (41502239, U1602233) and the Monitoring, Prediction and Research-combined Project of the China Earthquake Administration (163102).

## References

Bagnato, E., Aiuppa, A., Parello, F., D'alessandro, W., Allard, P., & Calabrese, S. (2009). Mercury concentration, speciation and budget in volcanic aquifers: Italy and Guadeloupe (Lesser Antilles). *Journal of Volcanology and Geothermal Research, 179*(1), 96–106.

Cai, Z. H. (1980). Earth tides and seismic waves recorded by the level fluctuations in a deep borehole at Wali near Beijing. *Acta Seismologica Sinica, 2*(2), 205–214. (**in Chinese**).

Campbell, W. H. (2009). Natural magnetic disturbance fields, not precursors, preceding the Loma Prieta earthquake. *Journal of Geophysical Research, 114*, A05307. https://doi.org/10.1029/2008JA013932.

Cao, X. L., Xue, J., Wang, H. S., Li, Y. C., & Li, Y. (2002). Dynamic tendency of underground water level and its imminent anomalies. *Earthquake, 22*(1), 97–103. (**in Chinese**).

Cao, X. L., & Zhang, S. X. (1999). Analysis of water level dynamics and anomalies of earthquake precursors in Wanquan well. *Acta Seismologica Sinica, 21*(3), 323–328. (**in Chinese**).

Che, Y. T., Liu, Y. W., & He, L. (2015). Hydrogen monitoring in fault zone soil gas-a new approach to short/immediate earthquake prediction. *Earthquake, 35*(4), 1–10. (**in Chinese**).

Che, Y. T., Liu, X. L., Yao, B. S., Yu, J. Z., Zhang, P. R., Liu, W. Z., et al. (2003). Well water temperature behaviors in capital circle and their formation mechanism. *Seismology and Geology, 25*(3), 403–420. (**in Chinese**).

Che, Y. T., Liu, C. L., & Yu, J. Z. (2008). Micro-behavior of well-water temperature and its mechanism. *Earthquake, 28*(4), 20–28. (**in Chinese**).

Che, Y. T., & Yu, J. Z. (2006). *Underground fluids and earthquake*. Beijing: China Meteorological Press. (**in Chinese**).

Che, Y. T., Yu, J. Z., & Liu, C. L. (1997). The hydrodynamic mechanism of water radon anomaly. *Seismology and Geology, 19*(4), 353–357. (**in Chinese**).

Che, Y. T., Yu, J. Z., Liu, C. L., & Sun, T. L. (2006). Study on influence distance of some interference sources around the station to the behavior of underground fluid. *Recent Developments in World Seismology, 4*, 10–16. (**in Chinese**).

Che, Y. T., Yu, J. Z., & Wu, J. Y. (1990). Study of the air-pressure effect on the deep well groundwater level in China. *Hydrogeology and Engineering Geology, 4*, 12–17. (**in Chinese**).

Chen, F. (1996). Hydrogen-the important source of fluid in earth interior. *Earth Science Frontiers, 3*(3–4), 72–79. (**in Chinese**).

Chen Y. J., Yang, X. X., & Liu, Y. M. (1994). Study of precise geotemperature observation in Capital Circle. In Department of Earthquake Monitoring and Prediction, China Seismological Bureau. *Observation and study of new methods for short imminent term earthquake prediction in Capital Circle* (pp. 134–140). Beijing: Seismological Press. (**in Chinese**).

Cheng, J. J. (1997). Elementary analysis of relationship between mercury anomaly of fault product gas and active fault. *Crustal Deformation and Earthquake, 17*(2), 97–100. (**in Chinese**).

Chi, Q. H. (2004). Abundance of mercury in crust, rocks and loose sediments. *Geochimica, 33*(6), 641–648. (**in Chinese**).

Cicerone, R. D., Ebel, J. E., & Britton, J. (2009). A systematic compilation of earthquake precursors. *Tectonophysics, 476*(3), 371–396.

Cui, Y. J., Du, J. G., Zhang, D. H., & Sun, Y. T. (2012). Application of remote gas geochemistry in earthquake prediction. *Advance in Earth Science, 27*(10), 1173–1177. (**in Chinese**).

Ding, F. H., & Ha, Y. Y. (2012). Study of relationship between Fengzhen well water and rainfall. *South China Journal of Seismology, 32*(1), 35–40. (**in Chinese**).

Dogan, T., Mori, T., Tsunomori, F., & Notsu, K. (2007). *Soil $H_2$ and $CO_2$ surveys at several active faults in Japan terrestrial fluids, earthquakes and volcanoes: The Hiroshi Wakita* (Vol. 2, pp. 2449–2463). New York: Springer.

Du, P. R. (1991). Barometric pressure variations and their influence on the crust's deformations and water level in deep well. *Acta Geophysica Sinica, 34*(1), 73–82. (**in Chinese**).

Du, L. T., Chen, A. F., Wang, J., & Huang, S. T. (1995). Earth's hydrogen removal. *Bulletin of Mineralogy, Petrology and Geochemistry, 3,* 193–195. **(in Chinese)**.

Fan, X. F., Huang, C. L., Liu, G. J., & Wang, X. Y. (2012). Primary analysis on observation data of trace hydrogen at Xiaxian, Shanxi. *Earthquake Research in Shanxi, 3,* 7–12. **(in Chinese)**.

Fang, Z., Liu, Y. W., Yang, X. H., Yang, D. X., & Zhang, L. (2012). Advance in the research on the source and migration mechanisms of gas in the seismic fault zone. *Progress in Geophysics, 27*(2), 483–495. **(in Chinese)**.

Fedotov, S. A. (1968). *On seismic cycle, opportunities of quantitative seismic regionalization and long-term seismic forecast* (pp. 121–150). Moscow: Seismic Regionalization in USSR.

Felzer, K. R., & Brodsky, E. E. (2006). Decay of aftershock density with distance indicates triggering by dynamic stress. *Nature, 441,* 735–738.

Field, E. H. (2007). Overview of the working group for the development of regional earthquake likelihood models (RELM). *Seismological Research Letters, 78,* 7–16.

Field, E. H., Dawson, T. E., Felzer, K. R., Frankel, A. D., Gupta, V., Jordan, T. H., et al. (2009). Uniform California earthquake rupture forecast, version 2 (UCERF 2). *Bulletin of the Seismological Society of America, 99,* 2053–2107.

Field, E. H., Jordan, T. H., & Cornell, C. A. (2003). OpenSHA: A developing community-modeling environment for seismic hazard analysis. *Seismological Research Letters, 74,* 198.

Fleischer, R. L. (1981). Dislocation model for radon response to distant earthquake. *Geophysical Research Letters, 8,* 477–480.

Freund, F. (1985). Conversion of dissolved "water" into molecular hydrogen and peroxy linkages. *Journal of Non-Crystalline Solids, 71*(1–3), 195–202.

Gao, Q. W., & Fan, S. Q. (1992). Abnormal reflecting of escaped gas in Xiadian fault before earthquakes. *Earthquake, 6,* 73–76. **(in Chinese)**.

Gold, T., & Soter, S. (1980). The deep-earth-gas hypothesis. *Scientific American, 242*(6), 154–161.

Gottardi, R., Kao, P. H., Saar, M. O., & Teyssier, C. (2013). Effects of permeability fields on fluid, heat, and oxygen isotope transport in extensional detachment systems. *Geochemistry, Geophysics, Geosystems, 14*(5), 1493–1522.

Guo, L. S., Liu, Y. W., Zhang, L., Liu, D. Y., & Li, Y. W. (2016). Present situation and future development of mercury precursory monitoring in China. *China Earthquake Engineering Journal, 38*(2), 303–308. **(in Chinese)**.

Hamada, K. (1993). Statistical evaluation of the SES predictions issued in Greece: Alarm and success rates. *Tectonophysics, 224,* 203–210.

Hamilton, R., & chair, the IDNDR Scientific and Technical Committee. (1997). *Report on early warning capabilities for geological hazards* (p. 35). Geneva: International Decade for Natural Disaster Reduction Secretariat.

Han, J., & Kamber, M. (2006). *Data mining: Concepts and techniques* (2nd ed., p. 800). San Francisco: Morgan Kaufmann Publishers.

Harris, R. A., Simpson, R. W., & Hermann, R. B. (1996). Stress relaxation shadows and the suppression of earthquakes: Some examples from California and their possible uses for earthquake hazard estimates. *Seismological Research Letters, 67,* 40.

Hauksson, E. (1981). Radon content of groundwater as an earthquake precursor: Evaluation of worldwide data and physical basis. *Journal of Geophysical Research, 86*(B10), 9397–9410.

He, G. Q., Chang, Q. J., Guo, Y. Y., & Song, Y. L. (1990). The dynamic variation of fault gases and earthquakes. *Northwestern Seismological Journal, 12*(4), 13–19. **(in Chinese)**.

Huang, F. Q., Chi, G. C., Xu, G. M., Jian, C. L., & Deng, Z. H. (2000). Research on the response anomalies of subsurface fluid in mainland monitoring network to the Nantou earthquake with $M_s$ 7.6. *Earthquake, 20*(Supplement), 119–125. **(in Chinese)**.

Huang, F. Q., Li, M., Ma, Y. C., Han, Y. Y., Tian, L., Yan, W., et al. (2017). Studies on earthquake precursors in China: A review for recent 50 years. *Geodesy and Geodynamics, 8*(1), 1–12.

Ingebritsen, S. E., & Manga, M. (2014). Earthquakes: Hydrogeochemical precursors. *Nature Geoscience, 7*(10), 697–698.

Johannesson, K. H., & Neumann, K. (2013). Geochemical cycling of mercury in a deep, confined aquifer: Insights from biogeochemical reactive transport modeling. *Geochimica et Cosmochimica Acta, 106,* 25–43.

King, C. Y. (1980). Episodic radon changes in subsurface soil gas along active faults and possible relation to earthquakes. *Journal of Geophysical Research: Solid Earth, 85*(B6), 3065–3078.

King, C. Y. (1984). Impulsive radon emanation on a creeping segment of the San Andreas fault, California. *Pure and Applied Geophysics, 122*(2–4), 340–352.

King, C. Y., Zhang, W., & Zhang, Z. C. (2006). Earthquake-induced groundwater and gas changes. *Pure and Applied Geophysics, 163*(4), 633–645.

Kissin, I., & Grinevsky, A. (1990). Main features of hydrogeodynamic earthquake precursors. *Tectonophysics, 178*(2–4), 277–286.

Klusman, R. W., & Jaacks, J. A. (1987). Environmental influences upon mercury, radon and helium concentrations in soil gases at a site near Denver, Colorado. *Journal of Geochemical Exploration, 27*(3), 259–280.

Kuo, T., Fan, K., Kuochen, H., Han, Y., Chu, H., & Lee, Y. (2006). Anomalous decrease in groundwater radon before the Taiwan M 6.8 Chengkung earthquake. *Journal of environmental radioactivity, 88*(1), 101–106.

Lai, G. J. (2014). *The response characteristics and mechanism of groundwater level to barometric pressure and earth tides.* China Earthquake Administration: Institute of Geophysics. **(in Chinese)**.

Lai, G. J., Huang, F. Q., & Ge, H. K. (2011). Apparent permeability variation of underground water aquifer induced by an earthquake: A case of the Zhouzhi well and the 2008 Wenchuan earthquake. *Earthquake Science, 24*(5), 437–445.

Li, Y. C. (1995). Coseismic spring anomaly of deep-hole waterlevel in Sichuan and earthquake prediction. *Sichuan Dizhen, 3,* 32–35. **(in Chinese)**.

Li, C. H., Chen, Y. H., Tian, Z. J., & Wang, Y. L. (1990). Theory of the response of well-aquifer system to earth tides and its application. *Hydrogeology and Engineering Geology, 6,* 10–13. **(in Chinese)**.

Li, Y., Du, J. G., Wang, F. K., Zhou, X. C., Pan, X. D., & Wei, R. Q. (2009). Geochemical characteristics of soil gas in Yanqing-Huailai basin. *Acta Seismological Sinica, 31*(1), 82–91. **(in Chinese)**.

Li, Q., Fu, H., Mao, H. L., Zhu, R. H., & He, D. Q. (2017). Study on the relationship between abnormal water temperature and earthquake in Yunnan area. *Journal of Seismological Research, 40*(2), 233–240. **(in Chinese)**.

Li, H. B., Wang, H., Xu, Z. Q., Si, J. L., Pei, J. L., Li, T. F., et al. (2013). Characteristics of the fault-related rocks, fault zones and the principal slip zone in the Wenchuan earthquake fault scientific drilling project hole-1 (WFSD-1). *Tectonophysics, 584*, 23–42.

Li, Y. P., & Xu, Z. F. (1990). The migration and enrichment of radon and its impact on environment. *Guangdong Geology, 14*(2), 75–78. **(in Chinese)**.

Li, H. B., Xue, L., Brodsky, E. E., Mori, J. J., Fulton, P. M., Wang, H., et al. (2015). Long-term temperature records following the Mw 7.9 Wenchuan (China) earthquake are consistent with low friction. *Geology, 43*(2), 163–166.

Lin, Y. W., Wang, J. H., & Gao, S. S. (1998). A new measurement method for $CO_2$ in fault gas and prediction of Zhangbei-Shangyi M 6.2 earthquake. *Earthquake, 18*(4), 353–357. **(in Chinese)**.

Liu, C. L., Che, Y. T., Xu, Y. J., Shi, Z. M., & Yang, X. H. (2011). Abnormal interruption of water flow from an artesian well prior to 2008 Wenchuan earthquake. *Geodesy and Geodynamics, 2*(2), 53–57.

Liu, Y. W., Fan, S. H., & Cao, L. L. (1999). Relationship between the medium-short term anomalies of ground fluid and the seismicitiy parameter. *Earthquake, 19*(1), 19–25. **(in Chinese)**.

Liu, Y. W., & Ren, H. W. (2009). Preliminary analysis of the characteristics of post-seismic effect of radon after the Wenchuan 8.0 earthquake. *Earthquake, 29*(1), 121–131. **(in Chinese)**.

Liu, Y. W., Ren, H. W., Zhang, L., Fu, H., Sun, X. L., He, D. Q., et al. (2015). Underground fluid anomalies and the precursor mechanisms of the Ludian $M_s$ 6.5 earthquake. *Seismology and Geology, 37*(1), 307–318. **(in Chinese)**.

Liu, Y. W., Shi, J., Pan, S. X., & Cao, L. L. (2002). Discrimination and regional features of medium and short-term precursors of ground fluid in Sichuan-Yunnan region. *Earthquake, 22*(1), 17–24. **(in Chinese)**.

Liu, Y. W., Sun, X. L., Wang, S. Q., & Ren, H. W. (2008). Relationship of bore-hole water temperature anomaly and the 2007 Ning'er M 6.4 earthquake. *Journal of Seismological Research, 31*(4), 347–353. **(in Chinese)**.

Ma, Y. C. (2015). Earthquake-related temperature changes in two neighboring hot springs at Xiangcheng, China. *Geofluids, 16*(3), 434–439.

Ma, Y. C., Huang, F. Q., Xue, Y., & Wang, B. (2015a). Statistical analysis of long-term observation data of water temperature in springs and wells. *Technology for Earthquake Disaster Prevention, 10*(2), 367–377. **(in Chinese)**.

Ma, W., Kong, X., Kang, C., Zhong, X., Wu, H., Zhan, X., et al. (2015b). Research on the changes of the tidal force and the air temperature in the atmosphere of Lushan (china) $M_s$ 7.0 earthquake. *Thermal Science, 19*, 487–493.

McGee, K., Casadevall, T., Sato, M., et al. (1982). *Hydrogen gas monitoring at Long Valley Caldera, California* (pp. 1–12). Menlo Park: Geological Survey.

McGee, K., Sutton, A., Sato, M., et al. (1983). Correlation of hydrogen gas emissions and seismic activity at Long Valley Caldera, California. *Eos, Transactions, American Geophysical Union, 64*(45), 891.

Meng, G. K., He, K. M., Ban, T., & Jiao, D. C. (1997). Study on activity and segmentation of active fault using measurements of radon and mercury gases. *Earthquake Research in China, 13*(1), 43–51. **(in Chinese)**.

Mileti, D. S., & Darlington, J. D. (1997). The role of searching in shaping reactions to earthquake risk information. *Social Problems, 44*, 89–103.

Mogi, K. (1985). *Earthquake prediction*. Tokyo: Academic press.

Mogi, T., Tanaka, Y., Widarto, D. S., Arsadi, E. M., Puspito, N. T., Nagao, T., et al. (2000). Geoelectric potential difference monitoring in southern Sumatra, Indonesia-coseismic change. *Earth Planets and Space, 52*, 245–252.

Roeloffs, E. A. (1988). Hydrologic precursors to earthquakes: A review. *Pure and Applied Geophysics, 126*(2), 177–209.

Satake, H., Ohashi, M., & Hayashi, Y. (1984). Discharge of $H_2$ from the Atotsugawa and Ushikubi faults, Japan, and its relation to earthquakes. *Pure and Applied Geophysics, 122*(2–4), 185–193.

Sato, M., & McGee, K. A. (1981). Continuous monitoring of hydrogen on the south flank of Mount St. Helens. *US Geological Survey Professional Paper, 1250*, 209–219.

Sato, M., Sutton, A., & McGee, K. (1984). Anomalous hydrogen emissions from the San Andreas fault observed at the Cienega Winery, central California. *Pure and Applied Geophysics, 122*(2–4), 376–391.

Sato, M., Sutton, A., McGee, K., & Robinson, R. S. (1986). Monitoring of hydrogen along the San Andreas and Calaveras faults in central California in 1980–1984. *Journal of Geophysical Research, 91*(12), 12315–12326.

Shangguan, Z. G. (1989). A study on the origin of the fault gas in west Yunnan province. *Earthquake Research in China, 5*(2), 51–56. **(in Chinese)**.

Shi, J. Z., & Wang, J. R. (2000). Analysis on the characteristics of radon in underground water before Gonghe earthquake with $M_s$ 7.0. *Earthquake Research Plateau, 12*(1), 68–70. **(in Chinese)**.

Shi, Z. M., & Wang, G. C. (2013). Relationship between the Earth tidal factor and phase lag of groundwater levels in confined aquifers and the Wenchuan $M_s$ 8.0 earthquake of 2008. *Science China: Earth Sciences, 56*(10), 1132–1140. **(in Chinese)**.

Shi, Z. M., & Wang, G. C. (2015). Sustained groundwater level changes and permeability variation in a fault zone following the 12 May 2008, Mw 7.9 Wenchuan earthquake. *Hydrological Processes, 29*(12), 2659–2667.

Shi, Z. M., Wang, G. C., Liu, C. L., Mei, J. C., Wang, J. W., & Fang, H. N. (2013). Coseismic response of groundwater level in the Three Gorges well network and its relationship to aquifer parameters. *Chinese Science Bulletin, 58*(25), 3080–3087.

Shi, Y., & Zhang, W. (1993). The mechanism of earthquake prediction by radon: The relationship between radon and solid tide. *Acta Seismologica Sinica, 15*(1), 103–108. **(in Chinese)**.

Singh, S., Kumar, A., Bajwa, B. S., Mahajan, S., Kumar, V., & Dhar, S. (2010). Radon monitoring in soil gas and ground water for earthquake prediction studies in North West Himalayas, India. *Terrestrial Atmospheric and Oceanic Sciences, 21*(4), 685–695.

Sugisaki, R., & Sugiura, T. (1985). Geochemical indicator of tectonic stress resulting in an earthquake in central Japan, 1984. *Science, 229*(4719), 1261–1262.

Sun, X. L., Wang, G. C., Shao, Z. G., & Si, X. Y. (2016a). Geochemical characteristics of emergent gas and groundwater in Haiyuan fault zone. *Earth Science Frontiers, 23*(3), 140–150. **(in Chinese)**.

Sun, X. L., Wang, J., Xiang, Y., & Wang, Y. X. (2016b). Statistical characteristics of subsurface fluid precursors based on

*Earthquake Cases in China. Earthquake, 36*(4), 120–130. (**in Chinese**).

Sun, X. L., Xiang, Y., Shi, Z. M., & Wang, B. (2017). *Preseismic changes of water temperature in the Yushu Well.* Western China: Pure and Applied Geophysics. https://doi.org/10.1007/s00024-017-1579-x.

Tao, S. F., Wang, C. L., Liu, Y. W., Jiang, D. Y., Shen, K. J., Ning, Z. Z., et al. (1992). A preliminary experimental study on interference factors of soil gas. *Northwestern Seismological Journal, 14*(1), 92–95. (**in Chinese**).

Tian, Z. J., & Gu, Y. Z. (1985). Analysis and processing of data on fluctuations of groundwater level. *Seismology and Geology, 7*(3), 51–59. (**in Chinese**).

Toutain, J. P., & Baubron, J. C. (1999). Gas geochemistry and seismotectonics: A review. *Tectonophysics, 304*(1), 1–27.

Wakita, H., Nakamura, Y., Kita, I., Fujii, N., & Notsu, K. (1980). Hydrogen release: New indicator of fault activity. *Science, 210*(4466), 188–190.

Wan, D. K., Wang, C. M., Li, J. C., Wan, D. B., Xu, X. L., & Huang, B. D. (1993). *Groundwater dynamic anomaly and short term earthquake prediction* (pp. 1–224). Beijing: Seismological Press.

Wang, D. (2000). Study on coseismic, postseismic effects of groundwater parameters in Tianshan seismically active area. *Inland Earthquake, 14*(3), 252–259. (**in Chinese**).

Wang, D. (2004). The dynamic abnormal variation of underground fluid in the north Tian Shan mountain before and after earthquakes with more than $M_s$ 7 in Xinjiang and its adjacent areas. *Inland Earthquake, 18*(1), 45–55. (**in Chinese**).

Wang, X. B., Chen, J. F., Xu, S., Yang, H., Xue, X. F., & Wang, W. Y. (1994). Geochemical characteristics of gases from hot spring in seismic region. *Science in China (Series B), 37*(2), 242–249.

Wang, C. Y., Du, J. G., & Zhou, X. C. (2004). Geochemical feature of mercury across sanhe-pinggu active fault. *Earthquake, 24*(1), 132–136. (**in Chinese**).

Wang, B., Huang, F. Q., & Jian, C. L. (2011). The influencing factors of radon from the Jiayuguan Zone and its earthquake reflecting effect. *Earthquake Research in China, 25*(3), 358–369.

Wang, C. M., & Li, X. H. (1991). Applications of fracture-gas measurement to the earthquake studies in China. *Earthquake Research in China, 7*(2), 19–30. (**in Chinese**).

Wang, B., Ma, Y. C., & Ma, Y. H. (2016). Variation of water temperature in the Yushu well and its correlation with the strong earthquakes in the Qinghai-Tibetan block. *Earthquake Research in China, 32*(3), 563–570. (**in Chinese**).

Wang, J. H., Zhang, P. R., & Sun, F. M. (1982). Another example of abnormal hydrogen evolution before and after earthquakes. *Earthquake, 4,* 17–34. (**in Chinese**).

Ware, R. H., Roecken, C., & Wyss, M. (1984). The detection and interpretation of hydrogen in fault gases. *Pure and Applied Geophysics, 122*(2–4), 392–402.

Wu, H. S., Bai, Y. S., Lin, Y. F., & Chang, G. L. (1997). The action of relay transmission of the radon migration. *Acta Geophysical Sinica, 40*(1), 136–142. (**in Chinese**).

Xie, T., Zheng, X. D., Kang, C. L., Ma, W. Y., & Lu, J. (2015). Possible thermal brightness temperature anomalies associated with the Lushan (China) *M* 7.0 earthquake on 20 April 2013. *Seismology and Geology, 37*(1), 149–161.

Xu, Z. Q., Li, H. B., & Wu, Z. L. (2008). Wenchuan earthquake and scientific drilling. *Acta Geologica Sinica, 82*(12), 1613–1622. (**in Chinese**).

Yan, R. (2008). *Study of several influence factor of well water level change.* China Earthquake Administration: Institute of Earthquake Science. (**in Chinese**).

Yan, R., Guan, Z. J., & Liu, Y. W. (2015). Hot spring water observations and its anomalies before the Lushan $M_s$ 7.0 earthquake in the western Sichuan region. *Acta Seismologica Sinica, 37*(2), 347–356. (**in Chinese**).

Yan, R., & Huang, F. Q. (2009). Preliminary study on coseismic response of Huanghua well water level to 5 times of the Sumatra earthquakes. *Earthquake Research in China, 25*(3), 325–332. (**in Chinese**).

Yan, R., Huang, F. Q., & Chen, Y. (2007). Application of wavelet decomposition to remove barometric and tidal response in borehole water level. *Earthquake Research in China, 23*(2), 204–210. (**in Chinese**).

Yang, M. B., Kang, Y. H., Zhang, Q., Bai, C. Q., Lin, Y. W., Wang, L., et al. (2009). Tendencious fall of groundwater table in Beijing region and recognition of earthquake precursor information. *Acta Seismologica Sinica, 31*(3), 282–289. (**in Chinese**).

Yang, T., Walia, V., Chyi, L., Fu, C., Chen, C. H., Liu, T., et al. (2005). Variations of soil radon and thoron concentrations in a fault zone and prospective earthquakes in SW Taiwan. *Radiation Measurements, 40*(2), 496–502.

Yu, J. Z., Che, Y. T., & Liu, W. Z. (1997). Preliminary study on hydrodynamic mechanism of microbehavior of water temperature in well. *Earthquake, 17*(4), 389–396. (**in Chinese**).

Yu, J. Z., Gu, Y. Z., & Yin, S. L. (1990). The discussion on the change of barometric coefficients in three wells and the relation with earthquake occuring. *Earthquake, 3,* 25–32. (**in Chinese**).

Zhang, Z. D., Chen, X. Z., Chen, J. M., Su, L. S., Wang, Z. M., Shi, R. H., et al. (1997). The short-term precursors research on earth tidal loading/unloading response ratio of groundwater-level. *Acta Seismologica Sinica, 19*(2), 174–180. (**in Chinese**).

Zhang, B., Fang, Z., Liu, Y. W., Yang, X. H., Zhao, G., & Jing, Y. (2014a). Relationship between water temperature anomaly and earthquake in Yunnan. *Earth Science-Journal of China University of Geosciences, 39*(12), 1880–1886. (**in Chinese**).

Zhang, X. D., Kang, C. L., Ma, W. Y., Ren, J., & Wang, Y. (2017). Study on thermal anomalies of earthquake process by using tidal-force and outgoing-longwave-radiation. *Thermal Science.* https://doi.org/10.2298/tsci161229153z.

Zhang, W., & King, C. Y. (1994). Radon measurement of the main fault zone in California, USA. *Acta Seismological Sinica, 16*(1), 118–123. (**in Chinese**).

Zhang, L., Liu, Y. W., Guo, L. S., Yang, D. X., Fang, Z., Chen, T., et al. (2014b). Isotope geochemistry of mercury and its relation to earthquake in the Wenchuan earthquake fault scientific drilling project hole-1 (WFSD-1). *Tectonophysics, 619,* 79–85.

Zhang, L., Liu, Y. W., Guo, L. S., & Zhang, G. M. (2015). Geochemistry of mercury in the fault zone. *Environmental Chemistry, 3,* 497–504. (**in Chinese**).

Zhang, W., Shen, C. S., Xing, Y. A., Wei, J. Z., Wu, R. G., Ju, H. P., et al. (1989a). A new index of short-term and imminent earthquake anomalies-content of mercury. *Earthquake Research in China, 5*(4), 13–19. (**in Chinese**).

Zhang, W., Wang, J. Y., & Er, X. M. (1988). *Principle and method of earthquake prediction by hydrochemistry.* Beijing: Educational Science Publishing House. (**in Chinese**).

Zhang, Y., Zhang, Z. B., Cao, X., Zhou, G. M., Guo, J., & Zhang, G. L. (2004). Analysis on the relationship between the earthquake and the escaping radon at Jiayuguan region as well as the anomalous characters of radon in groundwater before the Yumen earthquake. *Seismological and Geomagnetic Observation and Research, 25*(1), 78–82. **(in Chinese)**.

Zhang, Z. G., Zhang, S. X., Li, W., Yin, H. W., & Han, W. Y. (2007). Analysis on the mechanism of formation characteristics of water temperature tide in Changli well. *Earthquake, 27*(3), 34–40. **(in Chinese)**.

Zhang, Z. D., Zhen, J. H., & Feng, C. G. (1986a). Cubical dilatation effects of earth tide on water level observation in wells. *Journal of Seismological Research, 9*(4), 465–472. **(in Chinese)**.

Zhang, Z. D., Zhen, J. H., & Feng, C. G. (1986b). Effects of atmospheric pressure on observations of well water level variations. *Earthquake, 1,* 42–46. **(in Chinese)**.

Zhang, Z. D., Zhen, J. H., & Feng, C. G. (1989b). Quantitative relationship between barometric efficiency of well-water level and precipitation load efficiency. *Earthquake, 6,* 38–44. **(in Chinese)**.

Zhang, Z. D., Zhen, J. H., & Feng, C. G. (1989c). Quantitative relationship between the earth tide effect of well water level, the barometric pressure effect and the parameters of aquifers. *Northwestern Seismological Journal, 11*(3), 47–52. **(in Chinese)**.

Zhang, Z. D., Zhen, J. H., Zhang, G. C., & Jing, J. C. (1989d). Response of water level of confined well to dynamic process of barometric pressure. *Acta Geophysica Sinica, 32*(5), 539–549. **(in Chinese)**.

Zhao, G., Wang, J., He, A. H., Guo, M. X., Guo, B. L., & Qin, J. G. (2009). Study of normal geothermal dynamics. *Earthquake, 29*(3), 109–116. **(in Chinese)**.

Zhou, X. C., Du, J. G., Chen, Z., Cheng, J. W., Tang, Y., Yang, L. M., et al. (2010). Geochemistry of soil gas in the seismic fault zone produced by the Wenchuan $M_S$ 8.0 earthquake, southwestern China. *Geochemical transactions, 11*(1), 5–14.

Zhou, X. C., Wang, C. Y., Chai, C. Z., Si, X. Y., Lei, Q. Y., Li, Y., et al. (2011). The geochemical characteristics of soil gas in the southeastern part of Haiyuan fault. *Seismology and Geology, 33*(1), 123–231. **(in Chinese)**.

Zhu, Q. Z., Tian, Z. J., Zhang, Y. S., Gu, Y. Z., & Gao, S. S. (1997). Distortion of earth tide from water level in the Taipingzhuang well in Beijing and its relation to earthquakes. *Seismology and Geology, 19*(1), 50–52. **(in Chinese)**.

Zmazek, B., Todorovski, L., Džeroski, S., Vaupotič, J., & Kobal, I. (2003). Application of decision trees to the analysis of soil radon data for earthquake prediction. *Applied Radiation and Isotopes, 58*(6), 697–706.

(Received March 31, 2017, revised April 3, 2018, accepted April 27, 2018)

Printed in the United States
By Bookmasters